Collins

AF838194

White Rose Maths

White Rose Maths

Key Stage 3
Behind the Questions
Teacher Guide 2

Ian Davies and Caroline Hamilton

William Collins' dream of knowledge for all began with the publication of his first book in 1819. A self-educated mill worker, he not only enriched millions of lives, but also founded a flourishing publishing house. Today, staying true to this spirit, Collins books are packed with inspiration, innovation and practical expertise.
They place you at the centre of a world of possibility and give you exactly what you need to explore it.

Collins. Freedom to teach.

Published by Collins
An imprint of HarperCollins*Publishers*
The News Building
1 London Bridge Street
London
SE1 9GF

1st Floor, Watermarque Building
Ringsend Road
Dublin 4
Ireland

Browse the complete Collins catalogue at www.collins.co.uk

10 9 8 7 6 5 4 3

ISBN: 978-0-00-840092-7

British Library Cataloguing-in-Publication Data
A catalogue record for this publication is available from the British Library.

Authors: Ian Davies and Caroline Hamilton
Publisher: Katie Sergeant
Product Manager: Jennifer Hall
Product developer: Natasha Paul
Content editor: Tina Pietron
Editors: Karl Warsi, Julie Bond, Phil Gallagher, Tim Jackson, Eric Pradel
Proofreader: Laurice Suess
Project Manager: Karen Williams
Cover designer: Kneath Associates Ltd
Internal designer and illustrator: Ken Vail Graphic Design, Ltd
Typesetter: Ken Vail Graphic Design, Ltd
Production controller: Katharine Willard
Printed and Bound in the UK using 100% Renewable Electricity at CPI Group (UK) Ltd

Photo acknowledgements
The publishers wish to thank the following for permission to reproduce photographs. Every effort has been made to trace copyright holders and to obtain their permission for the use of copyright materials. The publishers will gladly receive any information enabling them to rectify any error or omission at the first opportunity.
p108 Rattanapon Ninlapoom/Shutterstock

FSC
www.fsc.org
MIX
Paper from responsible sources
FSC® C007454

This book is produced from independently certified FSC™ paper to ensure responsible forest management.

For more information visit: **www.harpercollins.co.uk/green**

Acknowledgements
The publishers gratefully acknowledge the permission granted to reproduce the copyright material in this book. Every effort has been made to trace copyright holders and to obtain their permission for the use of copyright material. The publishers will gladly receive any information enabling them to rectify any error or omission at the first opportunity.

Contents

Introduction: Behind the questions

The White Rose Maths approach

The White Rose Maths textbook series guides students and teachers through a comprehensive learning journey to help build a culture of deep understanding and competence in mathematics. It fits perfectly alongside the White Rose Maths Secondary schemes of learning but can also be used alongside other existing materials or curricula. The books support a mastery approach to teaching and learning and have been designed to support the aims and objectives of the Key Stage 3 National Curriculum. This states that all students should:

- become **fluent** in the fundamentals of mathematics, including through varied and frequent practice with increasingly complex problems over time, so that pupils develop conceptual understanding and the ability to recall and apply knowledge rapidly and accurately
- **reason** mathematically by following a line of enquiry, conjecturing relationships and generalisations, and developing an argument, justification or proof using mathematical language
- **solve problems** by applying their mathematics to a variety of routine and non-routine problems with increasing sophistication, including breaking down problems into a series of simpler steps and persevering in seeking solutions.

We recognise that time is an issue and it's not easy to cover the full content of the curriculum in the depth you would wish to. These books are designed to give time to think about a topic, to develop understanding and also to realise that mastery will not have been fully achieved by the end of a chapter, whatever its length. Our small step approach has been devised to ensure that students will return to a topic time and time again, both within the further study of that topic and also in other areas of mathematics; this iterative process of revisiting and interleaving will enable students to deepen their understanding.

A mastery approach

Helping students on their journey towards mastery is exactly what White Rose Maths exists to do. It isn't always an easy journey, and it's important to realise that there's more than one way to get there. We think mastery of anything – playing an instrument, speaking a new language, mathematics – takes a very long time. That's why we talk about the 'journey to mastery', rather than 'having mastered'. For example, children start learning to add in early years and keep developing their skills over many years – from single digit numbers, to multi-digit numbers, then decimals, then fractions, then negatives, addition in different units, and so on. It would take several years to fully master addition, arguably one of the most basic concepts in mathematics, so what we do is break the journey down into small steps, spending time carefully considering each step. Through intelligent practice and building up experience of different contexts, gradually we move towards mastery where students are fluent in the unfamiliar and can apply their skills in any new situation.

A concrete-pictorial-abstract approach is an important aspect of the mastery approach and is based on evidence and research. Some of this research comes from Jerome Bruner. In his research on the cognitive development of children (1966), he proposed three modes of representation:

- **Enactive representation (concrete)** – students should have the opportunity to use concrete objects and manipulatives to help them understand what they are doing.
- **Iconic representation (pictorial)** – alongside manipulatives, students can use pictorial representations. These representations can then be used to aid reasoning and problem solving.
- **Symbolic representation (abstract)** – both concrete and pictorial representations should support students' understanding of abstract methods.

The White Rose Maths textbook series is unique in that it has concrete and pictorial representations at its heart. Powerful tools such as bar models and double-sided counters will provide many lightbulb moments for your students.

Sequencing of content

Book 2 follows the same rationale as Book 1. Proportional reasoning is key to mathematical understanding, and the informal connections used in Book 1 are now formalised using the language of ratio and proportion. The first two blocks of Book 2 explore these ideas using the mastery approach, looking at concrete and pictorial representations alongside the abstract notation with colons. Once introduced, the ideas recur regularly throughout the book to help students to remember, and to deepen their understanding.

Much of the other content in this book also builds on students' learning in Book 1. It is not assumed that students will have remembered everything from their earlier studies and when necessary prior learning is reviewed before moving on to the newer content. Opportunities to revisit learning in other contexts are also included throughout the book. For example, ratio notation is included in questions in the percentages and angles blocks (amongst others) and algebraic notation and techniques are present in the vast majority of blocks.

The books are structured to support students in performing simpler tasks, so they can then move on to more complex tasks. For example, we cannot expect students to solve 'reverse percentage' before they understand what a percentage is. Take a look at the block overview on page 315.

The small steps progression from the White Rose Maths scheme of learning form the sequencing of concepts within each block. Take a look at this example from the Indices block.

As you can see, nothing is left to chance. The learning is broken down carefully to ensure that students have the appropriate building blocks to access all of the content. Before we ask students to explore and use the addition and subtraction rules of indices, we make sure they are confident with the meaning of expressions involving indices and can add and subtract like terms that involve indices. This helps prevent some of the more common misconceptions.

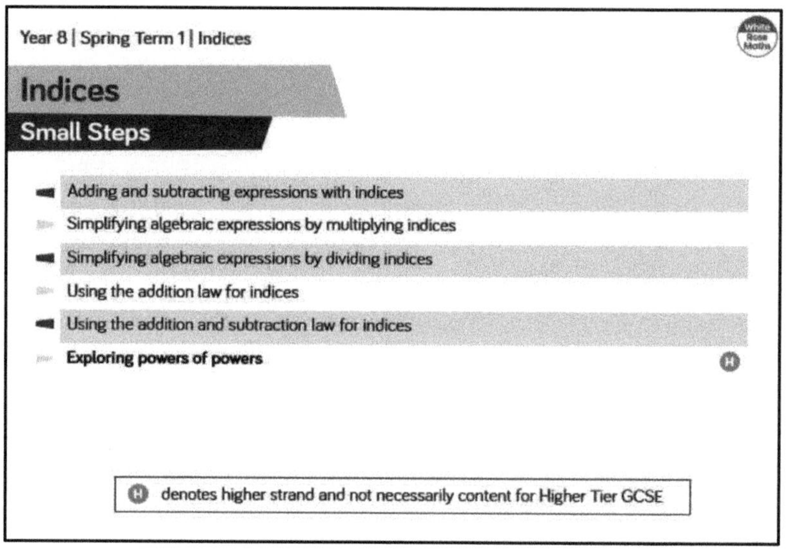

White Rose Maths Key Stage 3 textbook series components

Student book

There is one student book per year for KS3: Book 1, Book 2 and Book 3, corresponding to Years 7, 8 and 9 in the White Rose Maths schemes of learning. Each book caters for all attainment levels, and would work well for either setted or mixed-attainment groups. Each block of learning consists of:

- **A block opener** – This gives a visual of the topics that will be studied in the block and the key representations that will be used.
- **Between 2 and 8 chapters** – The number of chapters per block varies depending on the amount of content covered. Each chapter comprises:
  - A list of the White Rose Maths small steps that will be covered
  - The key words used in the chapter
  - An 'Are you ready?' section – this is a short set of questions to check that students possess the prior knowledge needed to access the new learning.
  - Visuals and accompanying explanation of the models and images used in the chapter, such as bar models or place value charts
  - Examples and practice – these are repeated in up to three sections: Practice A, Practice B and Practice C. The number of practice sections varies per chapter depending on how much content is being covered. Each practice section involves fluency, reasoning and problem-solving questions to ensure all students cover the main aims of the National Curriculum. Practice is often scaffolded with diagrams to help students to 'see the maths'. You will notice symbols are used throughout to prompt students to use concrete manipulatives, draw a diagram or discuss their thinking with others (these icons are fully explained in the Introduction at the start of the Student Book). These are all important aspects of a mastery approach to teaching and learning in mathematics.
  - 'What do you think?' sections – located at the end of each practice section, these questions are often more open-ended, or encourage deeper reasoning and problem-solving. We encourage all students to attempt these questions with support from whole class or paired discussion. Students should expect to have to think hard or think differently before answering.
  - A 'Consolidation' and a 'Stretch' section – towards the end of the chapter your class is likely to diverge. They will either complete the Consolidation section or the Stretch section depending on how they found the practice questions. Consolidation gives students a chance to attempt some more of the fluency-style questions. These are really useful if they have needed help throughout the main practice, as they could now attempt similar questions independently. Consolidation could also be used for revision, coming back to the topic at a later date, perhaps after assessment has suggested a gap in knowledge. Students who are confident with the practice questions get the chance to challenge themselves in the Stretch section. These questions often bring together numerous mathematical ideas and include rich problem-solving tasks.
  - A time to reflect – finally, students are asked to reflect on their learning, which may include putting key ideas into their own words, or justifying why something is or isn't true.
- **A 'Check my understanding' quiz** – Each block of learning ends with a short quiz covering content from all the chapters. Prior to attempting the quiz, students can read the statements about what they should have become fluent in and how they have used their reasoning and problem-solving skills. The questions in the quiz will indicate to both students and teachers the extent of retention of the key aspects of the block, and whether students need further revision or consolidation.

You will notice that we use an 'H' symbol at times throughout the student books. This denotes content to stretch and challenge students. It does not necessarily mean that it is higher content at GCSE; for example in Book 2, constructing bisectors is higher content but appears on the Foundation tier of GCSE. Use these chapters with as many students as possible, however do not worry if you miss them out. The higher content at

KS3 is covered again in later years when it might be more appropriate for some of your students. The 'R' symbol in Book 2 and 3 denotes review content. This material has been covered previously and this is a chance to revise and extend students' learning in these areas.

Behind the questions

A teacher guide is provided for each student book. The aim is to convey the rationale behind each chapter and give teachers an insight into the authors' thinking when formulating questions or sets of questions. You will find that conceptual and procedural variation is used throughout the books. The teacher guide helps you see when and why this is used.

Each Student Book chapter has a corresponding 'Behind the questions' section. This includes:

- The small steps covered from the White Rose Maths scheme of learning.
- The rationale for the learning taking place – why is it important, why is it taught in this place and what prior learning does it build on?
- Things to look out for – this section includes the common misconceptions or sticking points for students when learning this content.
- Interleaving opportunities – interleaving is an important aspect of successful learning and retention in mathematics. It is important that students see mathematics as an interconnected web of ideas and concepts, not stand-alone topics that have no relationship with each other. Interleaving not only helps students build connections but also gives the opportunity to revisit content and further consolidate it.
- Models and representations – a reminder of some of the key models and representations from the chapter and a rationale for why they are used.
- Commentary for the questions used in each section of the chapter – here, the authors choose some key questions and explain their thinking behind them. Often numbers will have been chosen very carefully to draw out a specific point. Reading this will help you, as the teacher, facilitate rich discussion and use careful questioning to draw out big ideas.

You can download editable versions of these teacher materials from collins.co.uk/WhiteRoseMaths/downloads

Additional materials and training available

White Rose Maths produce a great deal of extra material, often free, that would work well alongside this textbook series. They have a free, low-stakes assessment for each block of learning available plus a comprehensive set of more formal end-of-term assessments. These can all be accessed from their website www.whiterosemaths.com.

In addition, if you would like some training on some of the approaches used in the books, this can be booked either online or face-to-face. Top recommendations would be the courses entitled *Concrete-Pictorial-Abstract Methods*, *An Introduction to Bar Modelling*, *Mathematical Talk and Questioning*, *Reasoning and Problem Solving*, and *Thinking Through Variation*.

A note from the series editors

We are extremely excited by the release of this textbook series. Our passion is to help as many children and young people as we can to enjoy and understand mathematics. We hope that you can see this passion come through in our carefully thought-through questions, our use of models and images to aid understanding, and our interesting and rich reasoning and problem-solving tasks.

To us this is more than a textbook, it is a book full of exciting mathematical connections ready to inspire and enthuse teachers, students and parents alike, and we are incredibly proud to have joined the Collins team in the journey of its creation. We hope you enjoy using it as much as we enjoyed writing it.

Ian Davies (Head of Curriculum, White Rose Maths) **Caroline Hamilton** (Head of White Rose Maths)

Overview of course

Here is a handy overview of the blocks covered in each of the books in the White Rose Maths Key Stage 3 scheme, showing how you could fit them into your yearly programme for mathematics teaching.

Term	Book 1	Book 2	Book 3
Autumn	1. Sequences	1. Ratio and scale	1. Straight line graphs
	2. Algebraic notation	2. Multiplicative change	2. Forming and solving equations
	3. Equality and equivalence	3. Multiplying and dividing fractions	3. Testing conjectures
	4. Place value and ordering	4. Working in the Cartesian plane	4. Three-dimensional shapes
	5. Fractions, decimals and percentages	5. Representing data	5. Constructions and congruency
		6. Tables and probability	
Spring	6. Addition and subtraction	7. Brackets, equations and inequalities	6. Numbers
	7. Multiplication and division	8. Sequences	7. Percentages
	8. Fractions and percentages of amounts	9. Indices	8. Maths and money
	9. Directed number	10. Fractions and percentages	9. Deduction
	10. Fractional thinking	11. Standard index form	10. Rotations and translations
		12. Number sense	11. Pythagoras' theorem
Summer	11. Construction and measuring	13. Angles in parallel lines and polygons	12. Enlargement and similarity
	12. Geometric reasoning	14. Area of trapezia and circles	13. Solving ratio and proportion problems
	13. Developing number sense	15. Line symmetry and reflection	14. Rates
	14. Sets and probability	16. The data handling cycle	15. Probability
	15. Primes and proof	17. Measures of location	16. Algebraic representation
			17. Preparation for KS4

1.1 Representing ratios

Small steps

- Understand the meaning and representation of ratio
- Understand and use ratio notation
- Compare ratios and related fractions

Rationale

This opening chapter is designed to secure a deep understanding of the concept of a ratio as a comparison of two or more items. There is no need to develop procedures of simplifying or sharing using ratios, as these will be dealt with in later chapters. A key emphasis to consider is the notion of "for every *m* … there are *n* …", relating this to use of the colon notation. To avoid possible misconceptions, examples of the type where $m \neq 1$ are included from the start. Students may of course notice that an arrangement showing 4 : 2 is related to 2 : 1, as will be explored formally in Chapter 1.4.

Once the notation is understood, the second half of the chapter focuses on comparing ratios to fractions, in particular to clarify that if two quantities are in the ratio 1 : 2, say, then it is not the case that the first quantity is half of the whole. Bar models, counters and cubes are used to explore the fractions that can be seen from a ratio, with students challenged to think about the difference between one item as a fraction of another and as a fraction of the whole.

Things to look out for

- Writing ratios in the wrong order – correct order is emphasised throughout the chapter
- Confusing ratios and related fractions – this is also a key focus, with bar models being the key representation to show that 1 : 2, say, relates to the fractions $\frac{1}{3}$ and $\frac{2}{3}$

Interleaving opportunities

- Understanding of fractions from Book 1 is crucial
- Ratio notation is essential throughout this block and will feature heavily in future blocks, notably Blocks 2 and 12
- Pie charts are revisited, and you could use relevant questions from Book 1 to generate more practice questions if needed
- Use of letters to stand for numbers is revisited

Models and representations

- Students will be familiar with bar models from many contexts in Book 1. In particular, the use of the bar model to represent fractions features heavily in this chapter in order to prepare for their use to support problem solving in the remainder of the block.
- Counters, cubes, etc. will be useful to model the situations in questions so students can see relationships easily.

Behind the questions

Are you ready?

The first two questions revisit the representation of a fraction using a bar model, as this is key to the block. The third question provides a reminder of when to use numerators and denominators to compare fractions. The last part of this question provides a good basis for class discussion as to whether using common denominators, decimals or percentages is the most useful approach.

Practice 1.1A

1 and **2** provide simple practice in using the colon notation, with question 2 including comparison of the parts to the whole as well as to each other.

3 and **4** extend to comparing with more than two quantities, with great care needed with order. Students who find this tricky could use counters to model the situations and move these into the required order, to help.

5 Students may need reminding of the notation for labelling line segments. Many students will have found the exercise easy and could be challenged to create more complex questions based on this one.

What do you think? 1.1A

1 The ratio 1 : 1 is seldom explored and can cause confusion. It is worth noting that any situation comparing *n* with *n* reduces to 1 : 1, and also looking at 1 : 1.1, etc.

2 and **4** give a taster of simplifying, which will follow in Chapter 1.4.

3 provides a reminder of algebraic notation. If students find algebra difficult, this question is still easily accessible using five counters for *p* and three counters for *q*.

5 is the starting point for a wider investigation using longer strips, more colours etc.

Practice 1.1B

1 tackles a common misconception head on.

2 to **5** could all be supported by bar models or counters.

What do you think? 1.1B

1 links back to comparing fractions with equal numerators.

2 and **3** extend Practice 1.1B by looking at fractions of each other as well as the whole. Drawing diagrams "stacked" as in question 2 will provide support.

4 can be approached through trial and improvement and is good for comparing different answers.

Consolidate – do you need more?

This provides similar questions for further practice now or for revisiting at a later date.

Stretch – can you deepen your learning?

1 prepares for Block 2 by looking at ratio and multiplicative relationships.

2 tackles a common misconception around equivalence of ratios by comparing fractions.

3 and **4** use algebra to explore the ratio–fraction link more deeply and provide a good starting point for class discussion.

Reflect

Students' choice of representations here may be the basis for a display on ratio.

1.2 Solving ratio problems

Small steps

- Solve problems involving ratios of the form $1 : n$ (or $n : 1$)
- Solve proportion problems involving the ratio $m : n$

Rationale

Before moving on to dividing a quantity in a ratio in Chapter 1.3, working out one share given another is explored here, as this reinforces the idea of comparing which was covered in Chapter 1.1. The chapter starts by looking at what can be found out when a single "part" is known, exploring both the forms $1 : n$ and $n : 1$. Three-part ratios $1 : p : q$ or variants of this are also included.

Comparison bar models are the key vehicle used to represent problems and point the way to the solution. The same approach is recommended in the second part of the chapter with problems involving ratios of the form $m : n$, where both m and n are greater than 1.

Things to look out for

- Multiplying instead of dividing, and vice versa – encourage students to use labelled bar models to determine which operation to use. They should also check whether their answers make sense in the context of the question by asking themselves questions like "If the ratio is 1 : 5, which quantity must be greater?"
- Some questions include numbers which are divisible by both m and n, so students will need to read the questions carefully
- Students may expect that all answers will be integers, and may need reassurance when this is not the case

Interleaving opportunities

- The ideas of Chapter 1.1 are built on here
- This chapter links closely to all the chapters in Block 2, especially to Chapter 2.3 where both ratios and multipliers are considered in dealing with similar shapes
- Questions of the kind studied here are deliberately included in Chapter 1.3, to make sure students are thinking about the situation and not leaping to apply a procedure
- After completing Chapter 1.4, students could revisit their answers to this section to verify that they simplify to the ratios given in the questions
- Perimeter is revisited, and students may need to be reminded what this means

Models and representations

- The comparison bar model is key to the recommended approach to solving problems shown here. You will need to model its use carefully for students, emphasising the need to label both what is known and what they need to find out in order to take the correct first step. This is revisited in more detail in Chapter 1.3.
- The double number line is touched on as an alternative, and will be explored in greater detail in Block 2 as the link between ratios and multipliers is examined.

Behind the questions

Are you ready?

The first two questions revisit Chapter 1.1 to check that students understand ratio notation. Encourage them to draw bar models if necessary. The last two questions are precursors to finding other amounts when a unit is known or a multiple is known. Allow the use of calculators, as the focus need not be on arithmetic here.

Practice 1.2A

1 and **2** are scaffolded, with bar models provided, although some students may need help to label them. Encourage students to draw their own model for question 3.

4 uses the double number line as an alternative model. Use counters to represent the beads if needed.

5 can be supported using either model or by considering "for every 1 cm of width, there are 4 cm of length" and "for every 4 cm of length, there is 1 cm of width".

What do you think? 1.2A

1 encourages students to look for relationships on the double number line, preparing them for Block 2.

2 seems complex, but becomes much easier by drawing diagrams of the lines.

Practice 1.2B

1 and **2** suggest bar models and follow the pattern of the worked examples. You could discuss how, although these questions' contexts require integer solutions, there may be other contexts (e.g. lengths) that do not.

3 to **5** do not suggest bar models, but would clearly benefit from this approach.

What do you think? 1.2B

1 encourages deeper thinking about how to use a bar model given similar information.

2 could be extended to explore area ratios in more detail, but this is not necessary at this stage.

Consolidate – do you need more?

This section provides similar questions for further practice now or for revisiting at a later date.

Stretch – can you deepen your learning?

1 and **2** Students may need reminding of the angle sum of a triangle, and the use of three-letter notation to describe angles.

3 revisits substitution after applying the skills of this chapter in an algebraic context.

4 could lead to discussion as to whether constant rates are appropriate in this real-life context.

Reflect

Students can explain their chosen methods in their own words, providing useful reference for revision.

1.3 Dividing in a given ratio

Small steps

- Divide a value into a given ratio
- Solve problems involving ratios of the form 1 : *n* (or *n* : 1)
- Solve proportional problems involving the ratio *m* : *n*

Rationale

Although this chapter focuses on dividing a given quantity in a ratio, it is important that students tackle Practice 1.3B, as this includes examples of the form seen in the previous chapter. Hopefully this will prevent students from approaching all questions in the same way (as in "add, divide and multiply") regardless of the information given.

The fact that a comparison bar model can be used for all these situations should be emphasised, alongside the need for careful labelling of the known and desired quantities. Note that although a single bar may be useful for sharing a quantity in the ratio 5 : 3, for example, this is far less helpful if, say, the difference is known.

Things to look out for	Interleaving opportunities
• Students expecting integer answers – include examples where this is not the case • Falling into the trap of always adding to find the total number of parts, even when this is unnecessary or incorrect – use Practice 1.3B, and encourage students to devise their own varied questions • Difficulty setting up the model – you could scaffold by providing models in the first instance and asking students to explain your choices, and then gradually reduce the level of support	• The similarities between the problems here and in Chapter 1.2 are made explicitly, and are well worth discussing • As with Chapter 1.2, students could revisit their answers to this section after completing Chapter 1.4, to verify that they simplify to the ratios given in the questions • Fractions of an amount are discussed as an alternative approach • Angles in a triangle are included as a context; angles in other polygons and angle rules could also be included

Models and representations

- The comparison bar model is again the key model. Question 1 in Practice 1.3B is good for a discussion around how to label the model, and you may wish to consider a similar scenario with students building their models from scratch. They should be encouraged to use these models throughout the chapter to support clear solutions.

Behind the questions

Are you ready?

The first three questions revisit Chapter 1.2 and can be supported through bar models. The last question provides practice at finding a fraction of a known quantity, which may or may not be used as an approach to solving the problems in Practice 1.3A. Either way, it is useful revision, and the link can be a point for discussion.

Practice 1.3A

1 and **2** provide ample practice at standard "Divide … in the ratio …" questions. The effect of changing the number or the parts, but sometimes not the total, is worth a discussion.

3 to **5** explore different contexts.

6 mixes ratio, fractions and percentages in a style similar to many GCSE questions. Comparing how students choose to set out the solution is useful here.

What do you think? 1.3A

1 tackles the misconception that you always need to find the total number of parts.

2 and **3** explore situations when solutions may or may not be integers. Question 3 in particular provides an interesting discussion as to what to do with the extra penny!

4 is a forerunner to simplifying ratios. Using counters may be useful as a support.

Practice 1.3B

1 is only completed through careful reading of the scenarios and looking at the labels, which are skills this question is designed to develop.

2 to **3** do not explicitly suggest bar models, but again would benefit from this approach.

4 becomes awkward when the number of shares reaches large numbers, and also looks again at the need to round when finding a solution.

5 and **6** provide slightly different contexts, requiring careful reading of the language used.

What do you think? 1.3B

1 needs students to be able to convert between litres and millilitres (ml).

2 is a slightly more complex situation, but should be accessible.

Consolidate – do you need more?

This section provides similar questions for further practice now or for revisiting at a later date.

Stretch – can you deepen your learning?

1 is an interesting investigation, the result of which can be proved using fairly simple algebra stemming from the ratio $n : n + 1 : n + 2$

2 encourages students to consider the link between ratio and fractions of an amount.

3 extends to fractions greater than 1.

4 develops the idea further, including algebraically.

Reflect

The reflection explicitly addresses the key theme of interpreting and approaching similar but different problems.

1.4 Simplifying ratios

Small steps

- Express ratios in their simplest integer form
- Express ratios in the form 1 : *n* (H)

Rationale

Now that students are comfortable with the idea of a ratio as a comparison, we move on to looking at equivalent ratios. Cubes, counters or other objects can be used to model equivalence, linking back to the idea of "for every … there are …" and illustrating the convention of the simplest form of a ratio being given as $a : b$ where both a and b are integers. Students will need to be confident with common factors and should be encouraged to look for efficient ways of simplifying.

As a Higher step, the chapter also explores the forms 1 : *n* and *n* : 1. This is accessible to most students provided they are confident with decimals. As this is a Higher step, some of the questions are more challenging.

Things to look out for

- Only simplifying by dividing repeatedly by 2, and not appreciating that 3 : 6, say, cannot be simplified in this way – include examples with factors of 3 and 5 from the start to mitigate this
- Looking only at the numbers and not the units – discuss examples such as comparing £2 and 50p, asking which is bigger and therefore why the ratio cannot be 1 : 25
- Giving a ratio in the form 1 : *n* when this is not appropriate – emphasise that unless stated otherwise, "simplest form" means simplest integer form, and use physical objects to illustrate why this is the case
- Mixing up 1 : *n* and *n* : 1 – use careful questioning, asking "Which part do we want to be 1? Which part do we need to divide by?"

Interleaving opportunities

- Students need to be confident with factors, and the notion of highest common factor is particularly helpful when simplifying
- You could include more examples that revisit unit conversion, and if appropriate extend this to include imperial conversions and comparisons
- Students could now revisit their answers to problems in Chapters 1.2 and 1.3 to verify that their answers simplify to the ratios given in the questions
- This will link to forthcoming work on scales and maps
- Link to percentages, e.g. exploring why 60% links to 3 : 2

Models and representations

- Physical objects are highly useful for showing that 3 : 12, say, is equivalent to 1 : 4, for example by noting that "for every blue cube, there are four green cubes". This is particularly effective for when neither number in the simplified ratio is 1. For example, when simplifying 9 : 12, you can still see that "for every three blue cubes, there are four green cubes".
- You can use objects to "scale up" a ratio as well as to simplify.

Behind the questions

Are you ready?

The first two questions look at simplifying fractions, as the process is identical to simplifying ratios. Question 3 reminds students of some basic unit conversions that may be needed for simplifying comparisons given in different units. Question 4 explicitly looks at the language of "factor" and "highest common factor" that will be used when simplifying.

Practice 1.4A

1 encourages physical practice to develop understanding rather than just dividing procedurally.

2 and **3** provide fluency practice. Calculators or times tables grids may help with division if needed.

4 links to percentages. Some students may need reminding that 100% represents the whole.

5 and **6** look at units. Support students to convert first before simplifying.

7 and **8** encourage deeper thinking, with question 8 linking to fractions and multipliers.

What do you think? 1.4A

1 and **2** link explicitly to the previous two chapters.

3 makes the link to simplifying fractions.

Practice 1.4B

1 and **2** are fluency practice. Students may need help with rounding as this is familiar from KS2 but was not revised in Book 1.

3 to **6** explore practical contexts.

What do you think? 1.4B

1 links to upcoming work on reciprocals.

2 contains a lot of verbal information, but can easily be represented by a bar model.

3 A context such as number of adults per child is a good example.

Consolidate – do you need more?

This focuses more on the core Practice 1.4A than the Higher step Practice 1.4B.

Stretch – can you deepen your learning?

1 asks students to reflect on the meaning of non-integer results in recipes. They could look for other similar examples.

2 Pythagoras' theorem is covered in Book 3. Here it is just included as an application of 1 : 1 ratios. Other triangles could be explored if you like.

3 links to algebra. Students could extend by looking at other relationships.

Reflect

This provides an opportunity to cement the link between simplifying ratios and fractions.

1.5 Ratios in circles and graphs

Small steps

- Understand π as the ratio between diameter and circumference
- Understand gradient of a line as a ratio (H)

Rationale

In this chapter, two other areas of mathematics that link to ratios are explored. Students often find π a mystifying notion, so here we emphasise the fact that π is the ratio of the circumference of a circle to its diameter, i.e. $\pi = \frac{C}{d}$, as well as using the standard formula $C = \pi d$. Students should be encouraged to use 3 as an estimate for π when performing calculations, knowing that it will give an underestimate. The numbers have been chosen to avoid difficult rounding as the focus is on the relationship here.

Gradient is introduced on coordinate axes in Block 4 and studied in depth in Book 3. A Higher step is included to link the ratio of vertical and horizontal distances to gradient. In this chapter, only positive gradients are considered.

Things to look out for	Interleaving opportunities
• Using radius instead of diameter in $C = \pi d$ – this is explored here and, although $C = 2\pi r$ is discussed, the focus is on using a single formula and choosing the numbers carefully • Calculators may give answers as multiples of π – it is worth discussing how to change the display on students' calculators to avoid this • Difficulty with rounding – choose appropriate numbers so that rounding is only required to the nearest integer • Thinking that gradient must be an integer, or always greater than 1 – compare lines with gradients of 4 and $\frac{1}{4}$, say, and discuss the differences	• Perimeters of other shapes could be discussed • $C = \pi d$ will be revisited as a vehicle to practise rounding in Block 12 • Gradient is explored in depth in Block 4

Models and representations

- Use dynamic geometry software to illustrate how the circumference of a circle changes with its diameter, but the ratio remains constant.
- Students could estimate the circumference of round objects by eye and use tape measures to check, comparing to calculated answers.
- Dynamic geometry software can also be used to vary the slope of a line, or students can use geoboards to show line segments and their gradients.

Behind the questions

Are you ready?

Students are reminded about the perimeter of rectilinear shapes. Rounding is included for you to assess students' knowledge, but this is not a focus of the chapter. Question 3 looks at the equivalence of constant ratios in preparation for π.

Practice 1.5A

1 explores finding the ratio of the circumference to the diameter.

2 and **3** provide fluency practice, with question 3 looking at a given radius rather than diameter.

4 and **5** focus on parts of circles. Students may need help in determining divisors and what needs to be added to find their perimeters, with question 6 making the link to perimeter explicit.

What do you think? 1.5A

1 and **2** explore other ratios when dealing with circles. Part c of question 2 discusses the alternative formula $C = 2\pi r$, giving students the opportunity to deduce the connection between the formulae.

Practice 1.5B

1 looks at given right-angled triangles to find the gradient of a diagonal line.

2 requires students to draw the triangles themselves. Encourage large triangles that align on the gridlines.

3 and **4** require students to draw lines from given gradients, using squared paper.

What do you think? 1.5B

1 looks at the difference between a horizontal line (zero gradient) and a vertical line (infinite gradient).

2 compares commonly used phrasings of the formula to calculate a gradient.

3 introduces the idea of a negative gradient.

Consolidate – do you need more?

Both topics are revisited with further fluency practice.

Stretch – can you deepen your learning?

1 looks at approximations for π. This could be interesting to revisit after Block 3.

2 explores using the ratio of circumference and diameter to find the diameter (and radius) in a practical context.

3 requires students to be comfortable with unit conversions.

4 explores how gradient is represented on road signs.

Reflect

The reflections emphasise the link to ratio rather than the use of the formulae.

2.1 Direct proportion

Small steps

- Solve problems involving direct proportion
- Explore direct proportion graphs (H)

Rationale

Although students have worked on proportional reasoning throughout Book 1 and also in Block 1 of this book, this is the first time they will formally discuss direct proportion. It is useful to include real-life examples such as buying *n* items at the same price, and it is interesting to discuss when this does not hold, for example when buying a multipack. Extensive use is made of the double number line, also used in Block 1, so students see the links between direct proportion and ratio. Although the unitary method is touched upon, it is not the focus as sometimes students become fixated on this and miss more obvious multiplicative relationships. Students could use calculators throughout as the focus is on relationships rather than arithmetic.

The fact that when one variable is zero, the other is also zero needs emphasising regardless of whether students cover the Higher step in the chapter, which links this to the graph passing through (0, 0). This Higher material is quite accessible and could be covered by most students, or you could return to it when looking at graphs of the form $y = kx$ in Chapter 4.2.

Things to look out for

- Confusion with multipliers – even with simple cases such as doubling and tripling, it is worth pointing out that you can perform a division to find the multiplier rather than relying on "spotting", which will make it easier when you have to divide for either very large integers or non-integer multiples
- Mixing up when to divide and when to multiply – encourage students to consider the sense of their answers and then decide which operation is appropriate, asking questions like "Is there more petrol or less petrol? Will the cost be more or less? Does your answer make sense?"

Interleaving opportunities

- The use of multipliers is key throughout this block, and is revisited in Block 10
- Using non-calculator methods for multipliers provides good context when studying multiplication and division of fractions in Block 3
- Direct proportion graphs are revisited in Chapter 4.2
- Proportional relationships are also considered in Block 12

Models and representations

- Students will be familiar with the double number line from Block 1.
- Model problems by selecting appropriate parts of the double number line rather than drawing a full line every time. This is shown in the worked examples.
- A "ratio table" presenting the same information as part of a double number line is also used. Encourage students to work with their favoured representation.

Behind the questions

Are you ready?

The first question reminds students of simplifying ratios using common factors, which is analogous to upcoming double number line work. Questions 2 and 3 look at simple cases of finding multipliers in contexts already met. Discussion about the connections between the answers in question 4 is useful to highlight where students currently are with proportional reasoning.

Practice 2.1A

1 and **2** provide practice at using a double number line to explore common relationships.

3 is good for discussion of methods. In the first part, students should see that multiplying by 3 is quicker than dividing by 4 and then multiplying by 12. You may need to model that part b can be approached by multiplying by 1.5 as well as by the more common "adding on half again".

4 to **6** are also useful for discussing when the unitary method is or isn't appropriate.

What do you think? 2.1A

1 and **2** relate the topic to real-life situations and challenge the assumption that all relationships are directly proportional.

3 explores the need for a common multiplier. You could compare to finding whether the ratios $p:q$ are equivalent.

Practice 2.1B

1 to **3** explore drawing and reading off from direct proportion graphs.

4, by contrast, looks at a case where the relationship is not directly proportional.

What do you think? 2.1B

1 Emily's claim is true provided it is a point in addition to (0, 0).

2 checks students have understood the features of a direct proportion graph.

Consolidate – do you need more?

This provides similar questions for further practice now or for revisiting at a later date.

Stretch – can you deepen your learning?

1 and **3** link the idea of direct proportion to other areas of mathematics that students should be familiar with.

2 can be tackled in a variety of ways, including graphically, and is useful for class discussion.

4 requires accurate measuring!

Reflect

The questions explore students' understanding as well as procedural fluency.

2.2 Conversions

Small steps

- Explore conversion graphs
- Convert between currencies

Rationale

The use of conversion graphs follows nicely from direct proportion graphs, but also works as a stand-alone if your students have not covered this Higher step. Limits for the graphs that students have to draw are suggested, but scales will depend on the paper you provide for this. You may wish to spend some time discussing scales, relating back to intervals on a number line from Book 1. Students should be encouraged to draw lines when reading off conversion graphs rather than working "by eye", as this is much more likely to be accurate.

The conversion work in the second part of the chapter is expected to be done with the support of a calculator. Looking at "live" exchange rates is useful, comparing offers from different providers, and "sell" and "buy" rates. This is a good opportunity to revisit and discuss estimation. Encourage students to notice that currency conversion is always directly proportional, but not all of the conversion graphs have this relationship.

Things to look out for	Interleaving opportunities
• Difficulty drawing graphs – support students with choosing appropriate scales, or provide pre-drawn axes if necessary • Misreading scales – students sometimes assume that all scales increase by 1 every time, so model several different examples to support them • Deciding when to multiply and when to divide in currency conversion – help students to "sense-check" first by asking e.g. "Will there be more pounds than dollars? What would two pounds be worth? Does your answer seem reasonable?" • Problems with rounding – you many need to revise this, but it will be covered later in the book so do not make it the focus of the chapter	• Link back to direct proportion graphs • Algebraic graphs are covered in Block 4 • Rounding and estimation are both revisited in Block 12, and conversion of currencies is a good context for examples here

Models and representations

- The graph itself is the key representation used in the first part of the chapter and should be the focus of your modelling and teaching.
- Conversions can be supported using any of the models from the previous chapter, such as the double number line and ratio table.

Behind the questions

Are you ready?

The first two questions check that students are familiar with graphical representation in the first quadrant. The other two explore rounding so students can deal with difficult conversion rates in the second part of the chapter. Formal rounding to decimal places is covered in Block 12, although students will have covered this in primary school.

Practice 2.2A

1 and **4** You may wish to provide students with a copy of the graphs to draw on. Use the last part of question 4 to emphasise that not all conversions imply direct proportion.

2 and **3** Provide prepared axes if necessary. Encourage students to use (0, 0) and a known point as far from the origin as possible, rather than a series of coordinates.

What do you think? 2.2A

1 reiterates the need for (0, 0) and a "distant" point.

2 reiterates the need for (0, 0) to be a point on a direct proportion graph.

3 links forward to the upcoming currency conversion. You could compare graphical and calculated answers here.

4 looks at finding values outside the range of the graph. Encourage students to make the link to the direct proportion techniques explored in the previous chapter.

Practice 2.2B

1 and **2** provide fluency practice, explicitly including the "missing zeros" on a calculator display in showing \$5.40 as \$5.4

3 links back to the double number line.

4 looks at estimation, which can you encourage throughout the exercise.

5 and **6** require slightly more complex reasoning.

What do you think? 2.2B

1 and **2** develop from questions 5 and 6 of Practice 2.2B, extending use of conversion skills to problems with a few more steps.

3 might be more challenging. Suggest looking at £100 or €100 in both countries.

Consolidate – do you need more?

This provides similar questions for further practice now or for revisiting at a later date.

Stretch – can you deepen your learning?

1 and **2** provide opportunities to investigate actual exchange rates.

Reflect

As a relatively stand-alone topic, it is useful for students to record their reflections here to help them to remember their learning.

2.3 Similar shapes

Small steps

- Explore relationships between similar shapes
- Understand scale factors as multiplicative representations

Rationale

Shapes are now used as a vehicle to revisit both ratio and direct proportion, and to reinforce the links between them. Students are introduced to this by drawing enlargements with integer scale factors in problems that become gradually more abstract, looking at non-integer scale factors and pairs of shapes that are represented rather than constructed. Links to ratio are emphasised, and testing for similarity by looking for constant ratios is encouraged throughout.

Things to look out for

- Not recognising similar shapes when they are in different orientations and/or not identifying corresponding sides – encourage students to draw both shapes in the same orientation so as to more easily recognise corresponding sides
- Confusion over whether to multiply or divide by the scale factor – encourage students to sense-check their answers by asking e.g. "Is the length of the line in shape B going to be greater than or less than the length in shape A? Which calculation will give you the result you expect?"
- Not recognising enlargements by scale factors less than 1 as enlargements – work with these from the very start by emphasising that if A to B is an enlargement by scale factor a, then B to A is an enlargement by scale factor $\frac{1}{a}$

Interleaving opportunities

- Ratio notation is revisited
- Calculations of this kind can be explored again when looking at multiplication and division of fractions in Block 3
- Construction of triangles and the concept of perimeter are revisited

Models and representations

- Use squared paper to draw and compare similar shapes, so students see the effect rather than just performing calculations.
- Display photographs and their enlargements, comparing relative sizes of objects. Demonstrate the distorting effect of adjusting the scale of only one side of a shape or a photograph.
- A double number line or a ratio table can be used to help calculate unknown sides.

Behind the questions

Are you ready?

The first two questions check whether students can perform the sorts of calculations required in the chapter fluently. Allow the use of a calculator if appropriate. The remaining questions are reminders of the linked concepts of ratios and simplifying fractions.

Practice 2.3A

1 allows students to explore performing enlargements. Include more examples if necessary.

2 links similarity to ratio.

3 provides students with the opportunity to create their own examples and non-examples.

4 gives fluency practice at finding missing sides in simple problems.

5 Students may need reminding how to construct a triangle.

6 increases the challenge by rotating the shapes.

What do you think? 2.3A

All four questions explore the meaning of similarity rather than the procedural aspect of calculation which is developed in the rest of the chapter. Students should draw and measure shapes as support if needed.

Practice 2.3B

1 provides practice at identifying corresponding sides in preparation for question 4.

2 is preparation for the later need to use a calculator.

3 looks at enlargement by scale factors greater than and less than 1.

4 uses the same shapes as in question 1 for fluency practice.

5 can be approached by looking for common ratios. Students may need help to see that the order could change.

What do you think? 2.3B

1 and **2** focus on the calculations used in finding unknown sides, looking at how these might work.

3 requires fluency in converting units.

Consolidate – do you need more?

This provides similar questions for further practice now or for revisiting at a later date, again starting from drawing and observing.

Stretch – can you deepen your learning?

1 develops deeper understanding of similarity. Encourage students to draw shapes to test their opinions.

2 introduces the idea of area ratios, which are studied in depth at KS4.

3 and **4**, likewise, introduce students to the idea of similar triangles; there is no need to study this formally here.

5 extends this thinking to a pair of similar quadrilaterals.

Reflect

Here students reflect on the effect of different sizes of scale factor and the formal mathematical meaning of similarity.

2.4 Scale diagrams and maps

Small steps

- Draw and interpret scale diagrams
- Interpret maps using scale factors and ratios

Rationale

Building on from students' knowledge of scale factors in the previous chapter, they now explore using large scales in constructing, reading and interpreting plans and maps. Creating their own scale diagrams is important in developing understanding, and in demonstrating that the larger the scale, the smaller the resulting diagram.

It is useful to use plans and maps of local areas when teaching, as this both sparks interest and allows students to practise estimation. Some students may well be unaware of how far the school is from the town centre or other local landmarks, for example.

Things to look out for

- Mixing up scales such as "1 cm stands for 2 m" and "2 cm stands for 1 m" – address this directly by using a pair of scales like this for the same object, and encourage students to start by thinking "How many … are there? So how many … do I need?". Students should also check answers to see if they make sense.
- Difficulty with large scales such as 1 : 100 000 – encourage students to work one step at a time
- Thinking that 1 : 100 000 is more detailed than 1 : 50 000 – include this misconception as a teaching point by working out the relative sizes of plans of objects. It may be more appropriate to use 1 : 5 and 1 : 10, say, to make the point and then extend to larger numbers.
- Believing that all diagrams are drawn to scale – use sketch diagrams and scale diagrams to show how the former can be misleading

Interleaving opportunities

- Using ratio and simplifying ratio from Block 1 are explicitly revisited
- Unit conversions will be revisited in Chapter 12.3
- Link using the scale of a plan or a map to the scale factor of enlargement in Chapter 2.3

Models and representations

- Use squared paper to draw and compare diagrams that represent the same object at different scales. Help students to notice that the resulting diagrams are mathematically similar.
- Use real maps of the local area, preferably with a variety of scales.
- Use plans of the school buildings or houses from estate agents, for example, noting whether these are to scale.

Behind the questions

Are you ready?

The first two questions check whether students can perform the sorts of calculations required in the chapter fluently. Allow the use of a calculator if appropriate. The remaining questions are reminders of the linked concepts of ratios and simplifying fractions.

Practice 2.4A

1 and **2** provide practice at using various scales for the same diagram. Students may need reminding that 1 m = 100 cm. The possible confusion of 1m for 2 with 2 cm for 1 m is explicitly addressed.

3 illustrates that you can use scales with any units.

4 looks at the integer form of 1 : n for larger n. Students should be familiar with this simplification from Chapter 1.4, but may need a reminder. At this stage only metres and centimetres are included, but you could extend by including kilometres.

5 looks at calculating lengths without a diagram being provided.

6 and **7** look at choosing an appropriate scale. Ask students to consider factors such as what fits and what level of detail needs to be seen.

What do you think? 2.4A

1 is a useful discussion and can be related to solving problems involving unknown angles, for example.

2 to **4** all relate to skills covered in the main exercise at a slightly deeper level. Support students to look back at the similar questions they have already done.

Practice 2.4B

1 to **3** provide fluency practice at using larger scales from maps that are provided. The maps are related and provide a good opportunity for discussion of the advantages and disadvantages of using different scales and issues around accuracy of measurement.

4 involves the same process but without a map.

5 looks at determining larger map ratios. Encourage students to work step-by-step through these.

What do you think? 2.4B

1 looks again at the level of detail provided by different scales. Remind students of the diagrams they drew in question 1 in Practice 2.4A.

2 extends to finding a scale. Again, students should work step-by-step through the conversions and simplifications involved.

3 touches again on area ratio, also included in the previous chapter. Again, this is more of a discussion point than an in-depth introduction to the topic.

Consolidate – do you need more?

This provides similar questions for further practice now or for revisiting at a later date, again starting from students making their own scale drawings.

Stretch – can you deepen your learning?

1 makes the link between using different scales and similarity. You could also ask students to find the scale factors of enlargement between all the diagrams.

2 involves decimal conversion of units.

3 and **4** are points for discussion.

Reflect

These questions help students to make the connections to earlier learning.

3.1 Multiplying fractions by integers

Small steps

- Represent multiplication of fractions
- Multiply a fraction by an integer

Rationale

The purpose of this chapter is to build an understanding of multiplication of fractions by integers. Students should understand multiplication as repeated addition and use this as a foundation for their knowledge. They should already be confident when adding fractions with the same denominator, and so should be able to see that $3 \times \frac{1}{5} = \frac{1}{5} + \frac{1}{5} + \frac{1}{5} = \frac{3}{5}$. Students need to be exposed to examples both where the answer is proper and where it is improper. When answers are improper fractions or not in their simplest form, students should be able to convert them to mixed numbers and/or simplify when necessary. Pictorial representations of fractions can help students to understand these abstract calculations and are used throughout for support.

Things to look out for

- Multiplying both the numerator and the denominator by the integer, e.g. writing $\frac{3}{5} \times 2 = \frac{6}{10}$
- Drawing bar models touching so that the diagram no longer represents the calculation but instead just an equivalent fraction, e.g. drawing two bar models touching that each represent $\frac{3}{5}$ and hence together represent $\frac{6}{10}$ rather than $2 \times \frac{3}{5}$ (this can lead to the misconception described above)
- Incorrectly converting between improper fractions and mixed numbers

Interleaving opportunities

- Converting between improper fractions and mixed numbers
- Adding fractions with the same denominator
- Solving equations
- Area
- Algebraic notation
- Scaling problems

Models and representations

- Bar models are used to represent multiplication of fractions by integers and help strengthen the link between multiplication and repeated addition.
- A number line is used to aid understanding of the scale and size of fractions and converting between improper fractions and mixed numbers.

Behind the questions

Are you ready?

These five questions provide an opportunity for students to practise the key skills needed in this chapter. The numbers chosen are simple, but as ever calculators could be used if arithmetic is likely to be a barrier. Question 3 looks at repeated addition of fractions and in each part students add three of the same fraction. Students should reflect on what's the same and what's different between the question parts. In parts a and b the denominator is the only real difference, in part c the numerator isn't 1, and in part d the three fractions sum to one whole. Questions 4 and 5 ensure that students can convert between mixed numbers and improper fractions as this will be used in the lesson. Pictorial representations could be encouraged to support understanding.

Practice 3.1A

1 enables students to explicitly recognise that multiplying a fraction by an integer is the same as adding that fraction a given number of times.

2 does not require students to actually work out the answer to each calculation. The focus instead is on understanding the representation. Students could be encouraged to formally reflect on what each number in their written calculation represents in terms of the diagram.

3 Now students are required to understand the representation and work out the answer to the calculation. In parts a and b, students should recognise that whilst the representations look different, they both represent ninths. Whilst part a is 4 lots of 1 ninth, part b is 4 lots of 2 ninths, and they should reflect on what's the same and what's different in these parts. Similarly, in parts c and d they are working with tenths. Here they get the same answer and should understand why that is. In part d, some students might see this as 6 lots of 1 tenth rather than 3 lots of 2 tenths, which is fine, and again the understanding of why the answer doesn't change is important.

4 looks at the idea of unitisation to avoid misconceptions later in the chapter. Students should understand that the denominator in these questions is the "verb" and so doesn't change. If you multiply a given number of sevenths, for example, by an integer, you will still have sevenths – just a different quantity of them.

5 allows students to consolidate their learning by working in the abstract. All of the calculations give an answer which is a proper fraction that can't be simplified so that students need to focus on the new skill only. Some students might choose to draw pictorial representations for earlier parts but should understand why, once they get to part c on the top row and part h on the middle row, it becomes a very inefficient way of working. Parts a to e and f to j are connected: the denominators in the fractions are the same and it's the numerator that changes. Students should recognise what effect this has on the answer.

6 and **7** introduce students to calculations where the answer is an improper fraction and needs to be simplified. Students need to transfer their knowledge from earlier in the chapter along with learning from earlier in the curriculum to answer these.

8 gets students to think about missing numbers; they should realise when they get to part d that they will first need to convert $5\frac{1}{7}$ to an improper fraction to find the missing number.

What do you think? 3.1A

These four questions get students to really think about the representation of calculations and their answers, rather than just focusing on working out the answers by rote. In question 1, students should understand why the common misconception of multiplying the denominator by the integer can't be correct. This can be built up using multiplication of integers to see why it is that when you multiply by anything other than 1, the answer is not equal to the starting number. In question 4, students will need to recognise that $\frac{1}{2}$ is equivalent to $\frac{2}{4}$ to help them to answer the question.

Consolidate – do you need more?

These questions revisit the key points covered in the previous exercise, i.e. representing and calculating fractions multiplied by integers. If more practice is needed, students could work in pairs to produce similar questions and challenge each other.

Stretch – can you deepen your learning?

1 and **2** Here students are given worded questions. Students could be encouraged to consider the questions working in centimetres and minutes respectively without the fractions and then compare their answers.

3 to **5** provide more challenging questions linking the content from the previous exercise with previous learning on algebra. In question 5, students should notice that the value of x increases by 9 each time and reflect on why this happens.

6 Students are encouraged to generalise how to multiply a fraction by an integer. This could be built up, if necessary, using ideas such as $3 \times \frac{2}{7} = \frac{3 \times 2}{7}$ so $3 \times \frac{2}{c} = \frac{3 \times 2}{c} = \frac{6}{c}$ so $3 \times \frac{b}{c} = \frac{3 \times b}{c} = \frac{3b}{c}$, etc.

Reflect

Students should explain their learning in their own words as this will clarify their thinking. They should be able to identify how the key features of a calculation are represented in a diagram and understand how slight changes would affect this.

3.2 Multiplying fractions by fractions

Small steps

- Represent multiplication of fractions
- Find the product of a pair of unit fractions
- Find the product of any pair of fractions

Rationale

The purpose of this chapter is to build an understanding of multiplication of fractions. Students should be familiar with calculating area using multiplication and use this as a foundation for their knowledge. They should know that the area of a unit square is 1 square unit, as 1 × 1 is equal to 1. They can extend this idea by dividing the square into equally sized rows and columns to represent multiplication of fractions. Whilst pictorial representations of fractions are used to introduce the idea, students should be encouraged to move away from this and begin to generalise. They should understand the limitations of the pictorial representation, but understand that it is there to fall back on should they ever forget how to perform the calculation or want to sense-check their answers.

Things to look out for	Interleaving opportunities
• Only multiplying the denominators, not the numerators • Always finding a common denominator – whilst this won't make the answer incorrect, it will increase the room for error in the calculation • Incorrectly simplifying answers • Not simplifying answers when asked to	• Representing fractions • Simplifying fractions • Area • Multiplication • Powers • Directed number • Substitution

Models and representations

- A unit square is used when introducing multiplication of fractions so that students can link multiplication of fractions to area. If the side length of the square is 1 and it is split into three equal rows, then the height of each row must be $\frac{1}{3}$. Similarly, if the square is split into two equal columns, then the width of each column is $\frac{1}{2}$. Shading the overlapping section shows that $\frac{1}{3} \times \frac{1}{2} = \frac{1}{6}$.

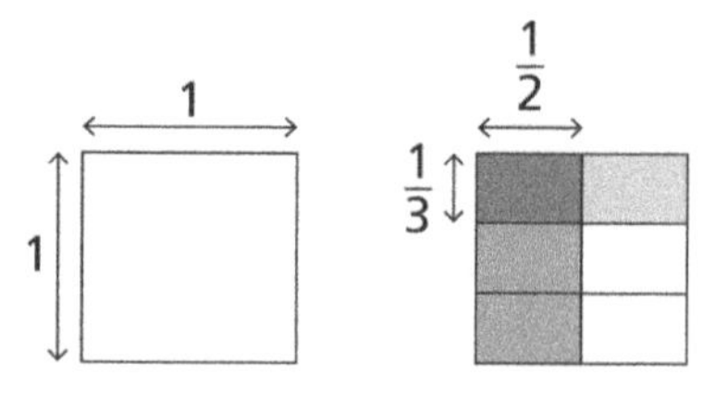

Behind the questions

Are you ready?

These five questions provide an opportunity to practise the key skills needed in this chapter. Question 1 recaps learning from the previous chapter whilst question 2 recaps students' understanding of unit and non-unit fractions. Throughout this chapter students will need to simplify their answers, so question 3 provides an opportunity for you to check they have retained this prior knowledge. Later in the chapter squaring and negative numbers are interleaved, so students are reminded of these skills in questions 4 and 5.

Practice 3.2A

1 uses the diagrams from question 1 in the "Are you ready?" section. This means that students should know the answer before they start, so the focus is on understanding the representation of the calculation and how this links to the answer rather than working out the answer.

3 encourages students to move away from the pictorial representation. Whilst it's a good way to develop conceptual understanding initially, students should understand the limitations and that it becomes quite time consuming as the numbers increase. It's important that they can explain their method clearly to demonstrate understanding of what's happening. Watch out for students saying "The numerator doesn't change" in their explanation. Although this will be more formally explored later in the chapter, it's better to avoid this misconception now.

6 Students could be asked to decide whether they think the order is important in part c, and this could be used to reinforce the idea of multiplication being commutative. In parts g and h, students should recall their knowledge of place value and understand that 0.1 is equivalent to $\frac{1}{10}$ and 0.01 is the same as $\frac{1}{100}$, and then use these facts to perform the calculations.

7 provides the opportunity to consolidate prior learning on directed number in the context of fractions.

What do you think? 3.2A

These two questions get students to think more deeply about multiplying unit fractions. In question 1, students should be able to explain in their own words why the numerator doesn't change when finding the product of a pair of unit fractions. In question 2, some students might be tempted to calculate fractions of an amount instead, but should be encouraged to think about it as a "multiplication of fractions" question.

Practice 3.2B

1 to **5** are similar to questions 2 to 5 in Practice 3.2A, but this time with non-unit fractions. Students could be asked to reflect on the similarities and differences between the corresponding questions in the two exercises, and to think about how these similarities and differences affect the answers.

6 Here students are instructed for the first time to give their answer in its simplest form. Students could be encouraged to consider whether it is more efficient to simplify before or after calculating. For example, in part g, they could either work out $\frac{8}{11} \times \frac{33}{80} = \frac{264}{880}$ and then simplify, or they could cross-cancel common factors and instead work out $\frac{1}{1} \times \frac{3}{10} = \frac{3}{10}$.

What do you think? 3.2B

These four questions get students to think deeply about multiplying non-unit fractions. In question 1, students are exposed to the misconception that the numerator doesn't change. In question 2, students are reminded of the idea of a fraction being one whole. They could link this to the idea of multiplying an integer by 1, whereby the number does not change, to help them to verbalise why the answer here is still $\frac{3}{4}$. In question 4, students should recognise that this answer must have been simplified, and work backwards to find the missing values.

Consolidate – do you need more?

These questions revisit the key points covered in the previous exercises. Question 1 gives shaded diagrams and question 2 gives diagrams to be shaded, but by question 3 students should be working in the abstract. Notice that it is not necessary to give answers in their simplest form in questions 1 and 2: the focus here should be on multiplying fractions, not simplifying. Simplifying is then revisited in question 3.

Stretch – can you deepen your learning?

3 Students often find multiplying integers more accessible than multiplying decimals. Since the calculations involved in multiplying fractions also involve multiplying integers, they should be encouraged to reflect on this and think of cases where fractions might be better than working with decimals.

4 Here students need to find the product of a pair of fractions, and then link back to learning from Book 1 on units of measure and fractions of an amount to answer part b.

5 Students often get confused with different rules when working with fractions. They should appreciate that although they can use a common denominator it isn't necessary in this case.

Reflect

Students should explain their learning in their own words as this will clarify their thinking. They should be able to identify how the key features of a calculation are represented in a diagram and understand how slight changes would affect this.

3.3 Dividing fractions (1)

Small steps

- Divide an integer by a fraction
- Divide a fraction by a unit fraction

Rationale

The purpose of this chapter is to introduce students to dividing fractions in a conceptual rather than a procedural way. Students will build on their knowledge of equivalent fractions – in particular, fractions that are equivalent to one whole – to support their learning. This chapter uses the idea of division being "how many lots of … there are in …". Although this sort of stem sentence is often easy to complete when working with integer values, it can feel a lot more abstract and challenging when working with fractions, so pictorial representations such as bar models and fraction walls can be used to support this.

Things to look out for	Interleaving opportunities
• Trying to move to a procedure too quickly with little understanding • Calculation errors which may be due to lack of understanding • When dividing an integer by a fraction, some students might divide by the denominator and multiply by the numerator (essentially multiplying the fractions instead of dividing)	• Representing fractions • Equivalent fractions • Multiplicative reasoning • Linear graphs • Solving equations • Substitution

Models and representations

- Bar models and fraction walls can be used to support understanding of dividing by fractions. These bar models show that there are 4 quarters in one whole, there are two lots of 2 quarters in one whole, and there are 3 quarters in three-quarters.

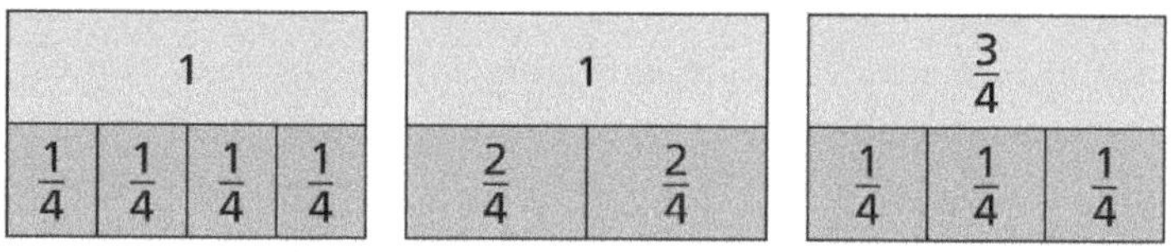

Behind the questions

Are you ready?

These five questions provide an opportunity for students to practise the key skills needed in this chapter. Question 3 reminds students about the actual meaning of division and how it is represented. Questions 4 and 5 recap equivalent fractions, with question 4 focusing specifically on fractions that are equivalent to one whole, as this is essential for the lesson.

Practice 3.3A

1 uses a bar model to support students' understanding of dividing an integer by a fraction. This could be linked to the idea that 1 is equivalent to $\frac{5}{5}$ to further support this. Students can often "see" the answer in part a by looking at the bar model but might find it more challenging to link the abstract calculations in part b to the diagram. This question links the abstract calculations to the question "How many … are there in …?"

3 Here the focus is not on the answer, and students are not required to it work out. Instead, the focus is on which calculations are equivalent and why. Students should explain their reasoning for each question: for example, 6 divided by one-fifth is equivalent to 6 multiplied by 5 because there are 5 fifths in one whole, and 6 wholes in 6. Parts a to d could be represented pictorially, if necessary, to support understanding, but students should be moving away from this and starting to work in the abstract by the time they get to part e, where the numbers increase and a pictorial representation would be time consuming.

4 Here the calculations are identical to those in question 3 but this time the focus is on working out the answer to the division. Students' answers to question 3 will support them with this.

5 starts to look at dividing an integer by a non-unit fraction. Students could copy the bar model into their books and highlight groups of two-fifths to support them with this if necessary.

8 provides students with practice at calculating abstractly. Notice the link between questions 4 and 8. Students have already worked out, for example, how many lots of one-fifth there are in 6 (for part a), so they can now look at how many lots of two-fifths there are. Students should focus on how their answers to question 4 are affected by the change in the numerator in this question.

What do you think? 3.3A

These three questions require students to look at dividing an integer by a fraction in a slightly different way. Although question 1 looks as though it is moving on to the procedural method of multiplying by the reciprocal, that isn't the case. Instead, the question is designed to link with the learning of the previous chapter: students should recognise that the calculations are the same, which is why they result in the same answer.

Practice 3.3B

1 uses a bar model to support students' understanding. This builds on the learning from earlier in the chapter and looks at dividing a fraction by a unit fraction.

3 uses a similar idea to question 1 in Practice 3.3A to support understanding. Students could be encouraged to shade or highlight a printed copy of the fraction wall to help them. As part of the discussion at the end of the question, students should be encouraged to think about the calculations that they would need to perform if they didn't have the fraction wall.

5 provides students with some abstract practice at dividing a fraction by a unit fraction. Notice how the equivalent fractions needed are the same as those given in the "Are you ready?" section. This means that students don't need to find any new equivalent fractions: instead, they can focus on mastering the new learning point, which is dividing a fraction by a unit fraction.

What do you think? 3.3B

These two questions require students to think more deeply about dividing fractions. In question 1, they should recall fraction equivalents from previous learning on place value. In question 2, students could be asked whether they actually need to work out the answers, or whether they can put them in order by just considering the calculation they would need to carry out.

Consolidate – do you need more?

These questions revisit the key points covered in the previous exercises. Question 1 focuses on eighths, whilst question 2 then builds on this to look at other denominators. Question 3 looks at different fractions given on the fraction wall. If any students require further practice, they could be encouraged to work with a partner to identify any other calculations they could work out using the fraction wall.

Stretch – can you deepen your learning?

1 to **3** These give students the opportunity to apply the new learning in contexts they've covered previously. Students should be encouraged to write down each calculation as well as just the answer.

4 encourages students to generalise their learning: for example, in part a, there must be b lots of b in one whole, so there must be a lots of b in a wholes. This could be scaffolded for students by suggesting values of a and b that they could use to work their way towards the generalisation.

Reflect

Students should explain their learning in their own words as this will clarify their thinking. They should be able to identify what steps they need to take to find the answer to a calculation and verbalise why they need to take them. It is important that they don't just remember a set of rules, as these are likely to become confused and are easy to forget.

3.4 Dividing fractions (2)

Small steps

- Understand and use the reciprocal
- Divide any pair of fractions

Rationale

The purpose of this chapter is to build on previous learning so that students can divide any pair of fractions. It's tempting when teaching this topic to teach a "trick" as it can make the work feel much easier, and students feel very successful at that moment. The issue with teaching it in this way is that students are less likely to remember the method if they don't have a conceptual understanding of the process. In the previous chapter, students divided integers by fractions and fractions by unit fractions, and they used the idea of the reciprocal without formalising it. This chapter formalises the idea of the reciprocal so that students can then apply it in more complex questions that are harder to represent pictorially. Developing an understanding of what is meant by "reciprocal" now will help students later in their education.

Things to look out for

- Trying to move to a procedure too quickly with little understanding
- Calculation errors which may be caused by lack of understanding
- Not simplifying fractions correctly
- Not converting their answer to a mixed number when asked to do so

Interleaving opportunities

- Representing fractions
- Equivalent fractions
- Mixed numbers and improper fractions
- Algebraic representation
- Solving equations
- Ratio

Models and representations

- Bar models and fraction walls can be used to support understanding of dividing by fractions. These bar models show that there are 4 quarters in one whole, there are two lots of 2 quarters in one whole, and there are 3 quarters in three-quarters.

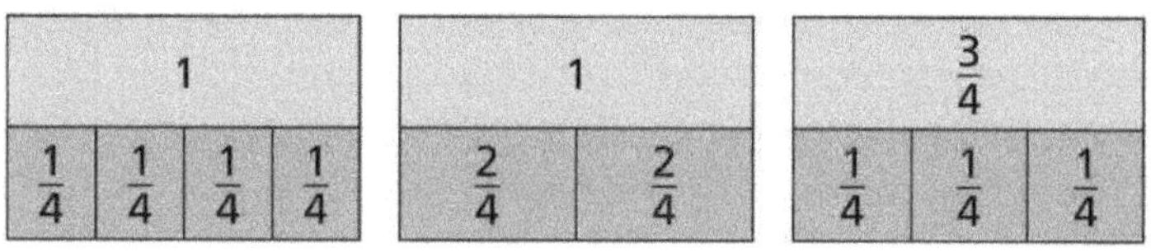

Behind the questions

Are you ready?

These five questions provide an opportunity for students to practise the key skills needed in this chapter. Question 1 looks specifically at calculations that give an answer of 1, which will support them when finding the reciprocal. Question 2 also supports the idea of finding the reciprocal, whilst question 3 uses the reciprocal without it being formalised. Both questions 3 and 4 revisit the content from the previous chapter, which students will build on in this one. Question 5 ensures that students can simplify fractions and convert from improper fractions to mixed numbers, as they will need to do this later in the chapter.

Practice 3.4A

1 provides the opportunity for students to look at reciprocals of unit fractions and integers. The question parts are linked to parts a to e of question 1 in the "Are you ready?" section so that students can look back at their answers to these for support. Parts v and vi highlight that if 5 is the reciprocal of $\frac{1}{5}$, then $\frac{1}{5}$ must be the reciprocal of 5.

2 Similarly to question 1, this allows students the opportunity to practise finding the reciprocal of non-unit fractions. Here this is broken down further to support understanding of making a product of 1 using fractions.

These link to parts f to j of question 1 in the "Are you ready?" section. As in question 1, parts v and vi are designed to highlight that if *a* is the reciprocal of *b*, then *b* is the reciprocal of *a*.

3 looks at content covered in the previous chapters but links it to the idea of the reciprocal so that students can generalise their learning and then start to apply it in more challenging contexts.

4 Here the focus is not on working out the calculations but, instead, students should think about what calculation would be equivalent. Discourage students from using language such as "You just flip it and times it". It's important that they understand that they are actually multiplying by the reciprocal. Most of the reciprocals required in this question have been worked out earlier in the exercise in questions 1 and 2, so that students don't need to focus too much attention on working these out: instead, they can focus on how knowing the reciprocal can help them to think about the calculation differently.

5 Now students are required to work out the answers to the divisions from question 4. In parts a to d they are required to divide an integer by a fraction. This is something they have done in a previous chapter but in a slightly different way. For example, for 5 divided by $\frac{3}{4}$, they will have previously thought of it as "There are 4 quarters in one whole, so there are 5 lots of 4 quarters in 5 wholes". Here they could be stretched to consider why the calculations involved in this exercise are identical to those seen previously. Students should be encouraged to consider the similarities and differences between parts e to h and parts i to l.

6 Here students are required to give their answers as mixed numbers in their simplest form. Many students can convert between mixed numbers and improper fractions and can divide fractions but struggle to combine the two. For students who are finding this difficult, you could help them by pointing out that all the answers have already been converted in the "Are you ready?" section.

7 Students should be encouraged to set out their working clearly and show all the steps to ensure that they are using inverse operations to solve the equations.

What do you think? 3.4A

These four questions require students to think about dividing fractions in slightly different ways. Question 1 uses reciprocals for support when dividing decimals. Question 2 revisits rules of negative numbers and asks students to think about this in the context of the reciprocal. Question 3 uses the rules of directed numbers along with substitution, and question 4 highlights three very different but valid methods that can be used to divide fractions. Students could be asked to consider the pros and cons of a "one method fits all" approach.

Consolidate – do you need more?

These questions revisit the key points covered in the previous exercise. Students are supported in finding the reciprocal and then required to use this to complete a division.

Stretch – can you deepen your learning?

1 and **2** look at the reciprocal in the contexts of algebra and decimals. In question 2, students need to recall and use their knowledge of place value and fractions.

3 looks at generalising division of fractions.

4 revisits ratio from earlier in the year.

5 and **6** look at conditions that affect the answer. Students will have seen throughout the previous exercises that sometimes the answer is an integer and sometimes it isn't; sometimes the answer is improper and sometimes it isn't. These two questions encourage them to think about when these types of answers occur and what it is that makes them occur. They could also be stretched further to consider when a fraction can be simplified and when it can't because it's already in its simplest form.

Reflect

Students should explain their learning in their own words as this will clarify their thinking. They should be able to give examples to support their reasoning.

3.5 Multiplying and dividing complex fractions (H)

Small steps

- Multiply and divide improper and mixed fractions (H)
- Multiply and divide algebraic fractions (H)

Rationale

The purpose of this chapter is to consolidate knowledge of multiplying and dividing fractions in more challenging contexts. Here students are required to multiply and divide improper fractions and mixed numbers. Students should be exposed to a range of methods to support their learning with this. Students then move on to look at multiplying and dividing algebraic fractions. This provides opportunities to revisit algebraic notation and simplifying expressions whilst still consolidating their new learning. When working through this topic, students could be prompted to consider why visual representations of multiplying and dividing fractions are no longer suitable now that they're working in more challenging and abstract contexts.

Things to look out for	Interleaving opportunities
• Calculating $3 \times \frac{1}{2} \times \frac{1}{5}$ for $3\frac{1}{2} \times \frac{1}{5}$ rather than realising that $3\frac{1}{2}$ is $3 + \frac{1}{2}$ • Incorrectly converting between improper fractions and mixed numbers • Just multiplying the integers and multiplying the fractional parts of two mixed numbers, and so missing out the other two terms	• Equivalent fractions • Partitioning numbers • Mixed numbers and improper fractions • Algebraic representation • Simplifying expressions • Ratio • Solving equations

Models and representations

- The grid method of multiplication can be used to support understanding when multiplying two mixed numbers. This grid shows that $3\frac{2}{5} \times 5\frac{1}{4}$ can be calculated by partitioning each mixed number so that the calculation becomes $(3 \times 5) + (3 \times \frac{1}{4}) + (\frac{2}{5} \times 5) + (\frac{2}{5} \times \frac{1}{4})$

	5	$\frac{1}{4}$
3	15	$\frac{3}{4}$
$\frac{2}{5}$	$\frac{10}{5}$	$\frac{2}{20}$

Behind the questions

Are you ready?

These four questions provide opportunities for students to practise the essential knowledge needed for this chapter. In question 1 they convert from improper fractions to mixed numbers and in question 2 they work the other way around, converting improper fractions to mixed numbers. It is a good idea to ask students to explain their methods to ensure that they understand the structure of a mixed number, as this will help them later in the lesson. Question 3 recaps learning from the previous chapters on multiplying and dividing fractions, whilst also offering the opportunity to consolidate simplifying fractions. Question 4 revisits learning on simplifying algebraic expressions, which will be used later in the chapter.

Practice 3.5A

1 looks at multiplying improper fractions and mixed numbers by integers. The mixed numbers used in parts a to d are equivalent to the improper fractions used in parts e to f. This allows students to compare the methods used when working with improper fractions and mixed numbers, and should prompt discussion on the pros and cons of each.

2 looks at the structure of a mixed number to eliminate the common mistake of just calculating $3 \times \frac{1}{5} \times \frac{1}{4}$. An area model could be used to support understanding in this question, and could then be used to find the answer to the actual calculation in part b.

3 provides the opportunity for students to consolidate multiplying a mixed number by a fraction. Encourage students to compare methods and explain why they do or do not convert their answers to improper fractions.

4 builds on learning from questions 2 and 3 to find the product of a pair of mixed numbers. Students could be stretched to consider when each method would be more appropriate.

5 allows students to consolidate their understanding of multiplying improper fractions and mixed numbers. Parts a and b could be done without converting between improper fractions and mixed numbers, but in parts c and d conversions are needed and students should explain their choice.

6 and **7** introduce the idea of dividing mixed numbers. In question 7, notice the links between parts a to d and parts e to h. If students already know how many fifths/halves/thirds/eighths there are in the mixed number, then challenge them to think about how their answers can be adapted for these questions.

8 looks at dividing fractions by first converting them to improper fractions. Students could be asked to consider why this is necessary when dividing mixed numbers.

What do you think? 3.5A

These two questions require students to look at multiplying and dividing improper fractions and mixed numbers more deeply. Question 1 highlights the reason why the two calculations are not the same and question 2 interleaves prior learning of decimals.

Practice 3.5B

1 to **4** look at multiplication and division of algebraic fractions. The level of difficulty builds up in question 1: the fractions don't need to be simplified until part d. In questions 2 to 4 the same expressions are used, and students should be encouraged to look at the connections between the questions and how these affect the answers. In question 4 students need to simplify by "cancelling" a letter from the top and bottom and might need prompting to recognise this.

5 gives students the chance to practise multiplying and dividing algebraic fractions where the answers need to be simplified.

What do you think? 3.5B

These three questions give students the opportunity to look at multiplying and dividing algebraic fractions more deeply. In question 3, students are likely to use their new "procedure": this is fine, but they should then be encouraged to think about how they could have answered the question more simply by spotting that the two fractions are equivalent.

Consolidate – do you need more?

These questions revisit the key points covered in the previous exercises. The focus here is not on simplifying the answers: instead, it is on consolidating the new learning point.

Stretch – can you deepen your learning?

1 and **2** look at multiplying and dividing complex fractions in the context of area. In question 1, students might need reminding that dividing by 2 is the same as multiplying by $\frac{1}{2}$ because $\frac{1}{2}$ is the reciprocal of 2.

4 Watch out for students not using the order of operations correctly or making mistakes when adding fractions because this hasn't been the focus of the learning during the previous chapters. Directed numbers are also used here and students should pay close attention to this.

5 Some students may not fully simplify their answer: encourage them to do so.

Reflect

Students should explain their learning in their own words as this will clarify their thinking. They should be able to demonstrate different methods and explain each step that they take in these calculations.

4.1 Points and lines

Small steps

- Work with coordinates in all four quadrants
- Identify and draw lines that are parallel to the axes

Rationale

Here students are reminded of coordinates and look at them in all four quadrants. They will have met coordinates in the first quadrant in Year 4 and will have seen coordinates in all four quadrants later in KS2. This step ensures that students are confident at plotting and reading coordinates in any quadrant. Once students have a deep understanding of what coordinates are and what they represent, they will transfer this knowledge to looking at equations of lines parallel to the axes. This is likely to be the first time they have worked with these lines. The use of correct mathematical language when reading the equation of a line can support understanding. For example, reading "$x = 4$" as "x is equal to 4" helps students to understand why at every point on this line the x value is equal to 4.

Things to look out for	Interleaving opportunities
• Writing coordinates the wrong way round • Getting confused when either the x or the y value in a pair of coordinates is equal to 0 • Not paying attention to the scale used on an axis • Drawing lines of the form "$x = \ldots$" parallel to the x-axis • Drawing lines of the form "$y = \ldots$" parallel to the y-axis	• Algebraic notation • Properties of shapes • Area and perimeter

Models and representations

- A coordinate grid is used throughout. Although not necessarily a model or a representation, having this visual aid supports students' understanding of which quadrant is which.

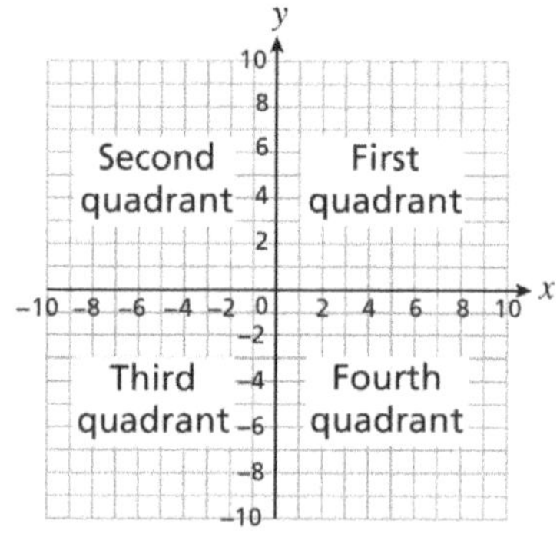

Behind the questions

Are you ready?

These four questions provide opportunities for students to consolidate the essential knowledge for this chapter. In questions 1 and 2, students demonstrate their ability to read and write coordinates in the first quadrant. Questions 3 and 4 are designed to ensure that students understand what each number in a coordinate pair represents, and which number is the x and which is the y value.

Practice 4.1A

1 and **2** ensure that students can transfer their knowledge from the "Are you ready?" section to working in four quadrants. Watch out for students making errors when working with points that lie on the axes. This is often a sticking point so there is plenty of practice of this here.

3 addresses two common misconceptions: writing the numbers the wrong way round, and counting squares rather than reading the values from the axes and thinking of the position relative to the origin.

4 looks at different scales to highlight the need to pay attention to the scale given.

5 interleaves knowledge of properties of shapes. Students should know straight away that the shape is a quadrilateral as it has four vertices, and they can plot these to see that it is a rectangle. Students could be stretched to think about how they could have realised that the shape was a rectangle without plotting the vertices.

What do you think? 4.1A

These three questions provide the opportunity to look at coordinates in all four quadrants more deeply. In question 1, students are asked to state the quadrant in which the points lie without plotting them. Question 2 looks at shapes drawn on a coordinate grid. Students should reflect on how they can find the side length of the square given the coordinates of the vertices. Question 3 focuses on the idea that when two points lie on the same horizontal/vertical line then they must have the same y/x value. This will support their learning in the next exercise.

Practice 4.1B

1 gets students to think about the x and y values in a pair of coordinates. In part a, the x value is equal to 4 and students should recognise that the y value can be anything: positive, negative, non-integer, etc. Then they consider more points that satisfy the same restriction. Students repeat this for y values.

2 Students build on learning from question 1 and draw some straight lines using this knowledge. They should soon realise that each set of points lie on a straight line and should be encouraged to explain verbally why any point that fits the given criterion will lie on the straight line.

3 and **4** look further at lines of the form $x = a$ and $y = b$, and students start to generalise their key characteristics, such as which axis each line is parallel to. It can feel unnatural to students that lines of the form $x = a$ are parallel to the y-axis and lines of the form $y = b$ are parallel to the x-axis, so this might need discussing formally.

5 allows students to practise finding equations of lines parallel to the axes.

What do you think? 4.1B

These three questions look in more depth at equations of lines parallel to the axes. Question 1 involves thinking about points of intersection. It is important for students to know that although they can plot the lines to find these points, they don't actually need to. In part c, students should demonstrate that they can find the coordinates of the points of intersection without plotting. Question 3 formally looks at the equations of the axes.

Consolidate – do you need more?

These questions revisit the key points covered in the previous exercises. If any students need further consolidation, they should be encouraged to work with a partner to create similar questions for each other to answer.

Stretch – can you deepen your learning?

1 Students should be encouraged to explain why there is only one pair of possible answers for this question, unlike the one in the earlier "What do you think?" section, where there were three pairs of possible answers.

2 Students recall knowledge of area in the context of coordinates. They could be encouraged to generalise and realise that if the third vertex was anywhere on the line $y = 9$ or $y = -1$, then the triangle would have an area of 20 square units.

3 looks at coordinates in a problem-solving context. Students should use the fact that any points on the same horizontal/vertical line have the same y/x value. If students need further support, they could draw the lines on a squared grid to aid understanding and then reflect on how they could have found the answer without this.

4 Although students have not formally covered lines of symmetry yet in KS3, they will have seen this in KS2, and this question gives them the opportunity to revisit this whilst also looking at equations of lines. Parts b and c have multiple answers, and students could be prompted to identify any conditions that these points need to satisfy in order for each shape to be correct.

Reflect

Students should explain their learning in their own words as this will clarify their thinking.

4.2 Lines of the form $y = kx$

Small steps

- Recognise and use the line $y = x$
- Recognise and use lines of the form $y = kx$
- Link $y = kx$ to direct proportion problems
- Explore the gradient of the line $y = kx$ (H)

Rationale

Here students are introduced to lines of the form $y = kx$ for the first time. The use of correct mathematical language when reading the equation of a line can support understanding. For example, reading "$y = x$" as "y is equal to x" helps students to understand why, at every point on this line, the y value is equal to the x value. Similarly, reading $y = 4x$ as "y is equal to 4 multiplied by x" helps students to understand why, at every point on the line, the y value is 4 times the x value. Once they have become familiar with lines of this form, students can then make the connection with direct proportion, and link the equation of a line to the context given in the question. Higher attaining students explore the gradient of the line $y = kx$ and the effect of increasing or decreasing the value of k.

Things to look out for	Interleaving opportunities
• Writing coordinates the wrong way round • Getting confused when either the x or the y value in a pair of coordinates is equal to 0 • Not paying attention to the scale used on an axis • Always drawing the line $y = x$ at a 45° angle to the axes regardless of the scales used • Getting the x and y values confused in a pair of coordinates	• Working with coordinates in all four quadrants • Lines parallel to the axes • Substitution • Algebraic notation • Direct proportion • Solving equations

Models and representations

- A coordinate grid is used throughout. It is important that close attention is paid to the scales used. Although all of these graphs appear to look the same, closer inspection reveals that they are not.
- A table of values will be used to work out coordinates of points that lie on a given line.

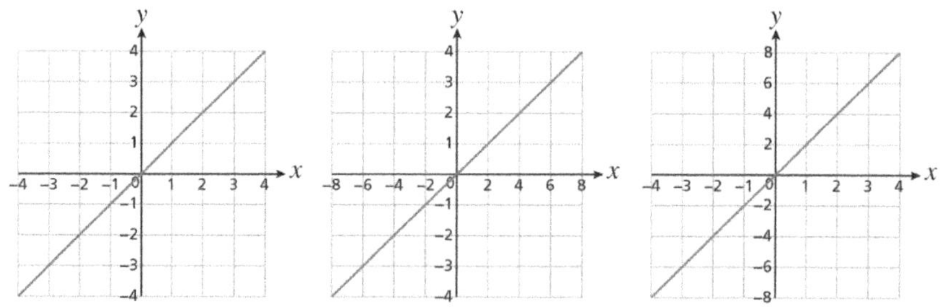

Behind the questions

Are you ready?

These four questions provide opportunities for students to consolidate the essential knowledge for this chapter. Questions 1 to 3 look back to the previous chapter to ensure that students understand how the equation of a line is related to the relationship between the coordinates at any point on the line. Question 4 involves substituting both positive and negative values into expressions of the form kx.

Practice 4.2A

1 to **4** Questions 1 and 3 ensure that students can link the relationship between the x and y values of any point on a given line and the equation of that line. Questions 2 and 4 then go on to look at completing a table of values and plotting the lines from questions 1 and 3 in a more formal way.

5 Students should be encouraged to think about the coordinates in the same way as they have done in the previous four questions. They should recognise that all of the lines go through the origin and could be prompted to explain why this happens.

What do you think? 4.2A

1 highlights the importance of the coordinates of any point on a linear graph lying in a straight line. The error made here is in relation to incorrect substitution of $x = 0$, but students should know that if their points do not lie in a straight line then they must have made a mistake.

2 addresses the common misconception that $y = x$ is always at 45° to the axes. Students should consider the similarities and differences between the three graphs.

3 By realising that each *y* value is 10 times the *x* value, this can be formalised to "*y* is equal to 10 times *x*" or $y = 10x$.

4 Students should know that when $x = 0$, the value of kx will be 0 for any given value of k, so lines of the form $y = kx$ must go through (0, 0).

Practice 4.2B

1 to **3** Question 1 focuses on integer numbers of pounds. Questions 2 and 3 move away from integer costs and do not specify whether to work in pounds or pence. Students could be encouraged to look at both units and compare the similarities and differences.

4 Students work in reverse and use the graph to find the information. They should be prompted to explain the relationship between the *x* and *y* values at any point on the line to support their reasoning.

Practice 4.2C

2 links the gradient to the sequence formed by the *y* values. This supports understanding of why a line is steeper when the value of *k* is greater. It means that, for every one unit across, the line is travelling further "up" and hence it is steeper.

3 generalises question 2. Watch out for students trying to be too exact and add scales to their graphs. It is important that they are able to sketch graphs and explain the key features of their sketches.

What do you think? 4.2C

Students could add scales to the axes that would make each line $y = x$, and also add scales that would make each line not $y = x$.

Consolidate – do you need more?

These questions revisit the key points covered in the previous exercises. If any students need further consolidation, they should be encouraged to work with a partner to create similar questions for each other to answer.

Stretch – can you deepen your learning?

1 Students should recognise that the line is of the form $y = kx$ because it goes through the origin. Then they can just use the relationship between 15 and 52.5 to enable them to find the equation of the line.

2 Watch out for students not recognising that different scales are used on each graph. They could be encouraged to find a different scale that would give a different equation.

3 Students use knowledge from earlier in the chapter to complete the scale on the *y*-axis, and then recap learning from earlier in the block to look at lines parallel to the axes.

4 and **5** further emphasise the relationship between points on a line and look at them in a slightly different way.

Reflect

Students should explain their learning in their own words as this will clarify their thinking. They should be able to use examples of lines and coordinates to support their reasoning.

4.3 Lines of the form $y = x + a$

Small steps

- Recognise and use lines of the form $y = x + a$

Rationale

Here students are introduced to lines of the form $y = x + a$ for the first time. As in the previous chapters, correct mathematical language when reading the equation of a line can support understanding. For example, reading "$y = x + 4$" as "y is equal to x plus 4" helps students to understand why, at every point on this line, the y value is 4 more than the x value. The focus here is on graphs that have a gradient of 1. So far in this block students have only seen equations with a coefficient of x. Now the narrow focus is on what happens when a number is added to or subtracted from x. Students should be encouraged to consider the similarities and differences between these lines and the line $y = x$. This will be built upon when looking at lines of the form $y = mx + c$ in Chapter 4.5.

Things to look out for	Interleaving opportunities
• Writing coordinates the wrong way round • Not paying attention to the scale used on an axis • Getting the x and y values confused in a pair of coordinates • Incorrect calculations with directed number • Not extending a line to the full width of the grid, but instead only drawing it between the points plotted	• Working with coordinates in all four quadrants • Lines parallel to the axes • Substitution • Algebraic notation • Solving equations • Directed number

Models and representations

- A coordinate grid is used throughout. It is important to pay attention to the scales used. Although all of these graphs appear to look the same, closer inspection reveals that they are not.
- A table of values will be used to work out coordinates of points that lie on a given line.

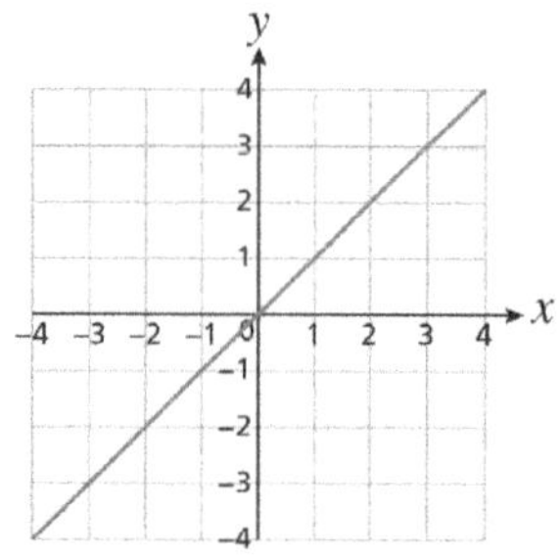

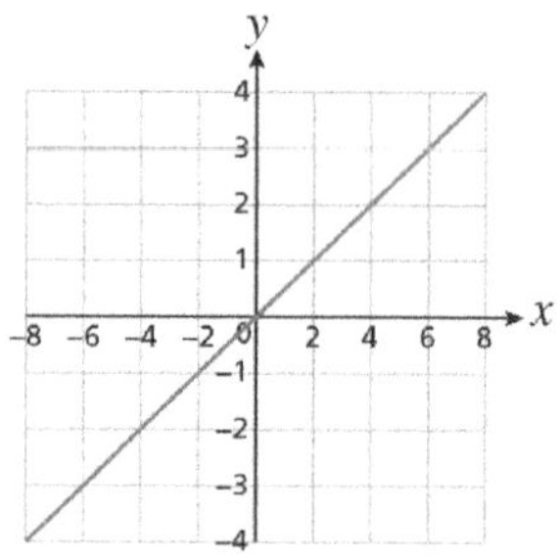

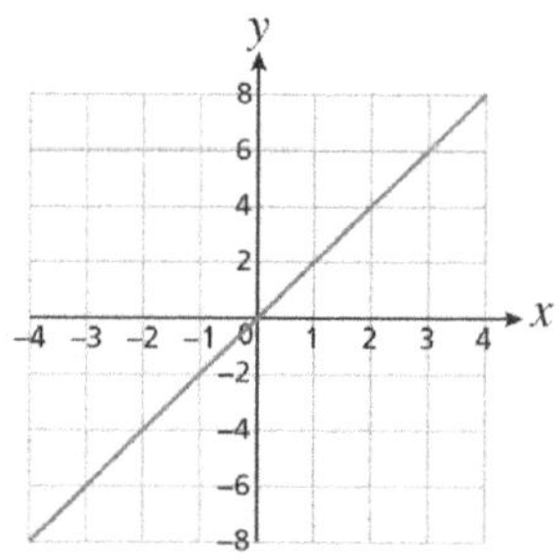

Behind the questions

Are you ready?

These four questions provide opportunities for students to consolidate the essential knowledge for this chapter. In question 1, students are required to describe an expression in words. This will help them to know what calculations to perform when substituting in questions 2 and 3. It will also support them when describing the relationship between the x and y values of any point on a line. In questions 2 and 3, students substitute a positive and a negative value of a, respectively. Notice how the expressions are the same in all of the first three questions. Question 4 then links these skills to coordinates of points on a line, and students should be encouraged to consider the similarities between parts a and b in questions 1 to 3 and this question.

Practice 4.3A

1 Students look at the relationship between the coordinates of any point on a line of the form $y = x + a$ for the first time. Describing the relationship in part a will make it easier for students to find the coordinates in part b. Students could compare their answers to part b to appreciate that there is an infinite number of possibilities.

2 Watch out for students making mistakes with directed number or confusing the x and y values.

3 Students now start to copy and complete tables of values and draw graphs of lines of the form $y = x + a$. These four graphs have been chosen to encourage students to spot patterns. Although they should notice that the lines are parallel, it is not important for them to know why at this stage. Instead, they should look at how the graph changes when the value of a changes. They do not need to use the language of "y-intercept" at this point.

4 uses the same approach as question 3 but for lines of the form $y = x + a$ where a is negative.

5 Now that students have seen lines of the form $y = x + a$ and have plotted them, they should be able to find the equation of a given line. Although some students might be able to use what they have learnt in questions 3 and 4 to just read off the equation, it is important that they can use coordinates to support their reasoning, and they should appreciate that if the scale was different then this might need more thought.

6 Here students could be encouraged to write down what they think is the equation of each line without any working, and then use working and coordinates to check their answers. They should reflect on the similarities and differences between line C and lines A and B.

What do you think? 4.3A

1 allows students to formally explore the differences between the line $y = x$ and lines of the form $y = x + a$. They should be able to explain, in their own words, the differences in the points on the lines and in what the lines look like when drawn.

2 Students often find lines presented in the form $y = a + x$ more challenging as this is not the way they are used to seeing the equation written. This question uses the commutative property of addition to help students to recognise that both equations are equivalent.

Consolidate – do you need more?

These questions revisit the key points covered in the previous exercise. Students use a function machine to support them in working out the coordinates of points on a line. They should be encouraged to explore how one question supports the next. If further practice is needed, students could be given the equations of more lines and asked to repeat these steps to plot them.

Stretch – can you deepen your learning?

1 Although they do not need to explore y-intercepts yet, this question gets students thinking about how you can tell from a diagram whether a line has a positive or a negative y-intercept. Understanding this will support them later in the curriculum.

2 Watch out for students not working out the scales on the axes and drawing a line parallel to the y-axis but through an incorrect point.

3 Students should be encouraged to draw a diagram to support their thinking. They should be able to explain each step of their working. Some confusion may arise because of the lack of units when calculating the area, but students should understand that just giving these as "square units" is sufficient in this context.

4 Students should understand that when a line intersects the x-axis, the y value at that point must be equal to 0. Spotting this is key to completing the rest of the question.

Reflect

Students should explain their learning in their own words as this will clarify their thinking. They should be able to use examples of lines and coordinates to support their reasoning.

4.4 Lines with a negative gradient

Small steps

- Explore graphs with negative gradient ($y = -kx$, $y = a - x$, $x + y = a$)

Rationale

Here students are introduced to lines with a negative gradient for the first time. So far in this block students have only seen lines with a positive gradient. As in the previous chapters, correct mathematical language when reading the equation of a line can support understanding. For example, reading "$y = -x$" as "y is equal to negative x" helps students to understand why, at every point on this line, the y value is the negative of the x value. Students should be encouraged to reflect on the similarities and differences between these lines and the ones they have seen previously.

Things to look out for

- Writing coordinates the wrong way round
- Not paying attention to the scale used on an axis
- Getting the x and y values confused in a pair of coordinates
- Incorrect calculations with directed number
- Not extending a line to the full width of the grid, but instead only drawing it between the points plotted
- When working with a line of the form $x + y = a$, completing a table of values by just adding a to x

Interleaving opportunities

- Working with coordinates in all four quadrants
- Lines parallel to the axes
- Substitution
- Algebraic notation
- Solving equations
- Directed number

Models and representations

- A coordinate grid is used throughout. It is important that close attention is paid to the scale used. Although all of these graphs appear to look the same, closer inspection reveals that they are not.

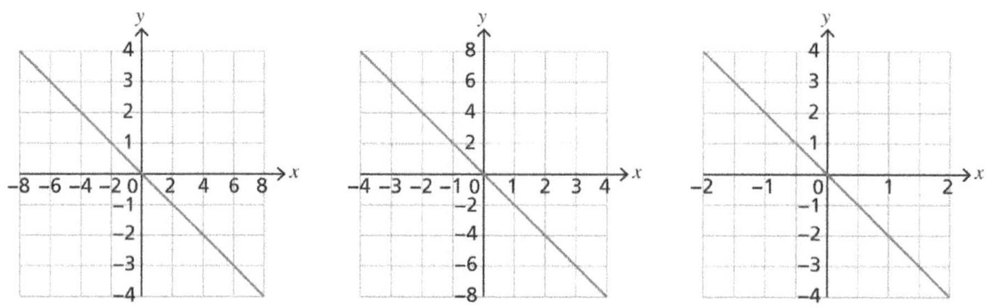

- A table of values will be used to work out coordinates of points that lie on a given line.

Behind the questions

Are you ready?

These three questions cover the essential prerequisite knowledge for this chapter. Question 1 provides practice at substituting into expressions as students will be required to substitute values into equations of lines in this chapter. Question 2 ensures that students can solve one-step equations as they will need to be able to find pairs of coordinates on lines of the form $x + y = a$. Question 3 asks students to complete a table of values for a line with a positive gradient.

Practice 4.4A

1 Students plot and look at the graphs of $y = x$ and $y = -x$ and compare the similarities and differences. It is important that they can recognise the key feature of a graph with a negative gradient, i.e. that it is sloping downwards. This will not only give them confidence when drawing graphs with a negative gradient, but also support their understanding of gradients later in the curriculum.

2 and **3** ask students to plot lines with negative gradients where the equation of the line is given in different forms. They should recognise that all of these lines slope downwards because their gradients are negative and could then be stretched to consider how they can tell from an equation that a line will have a negative gradient.

4 and **5** Students should now be able to recognise when a line has a negative gradient both from a graph and from an equation. Students often make mistakes with directed numbers, which result in lines being drawn with the wrong gradients. If students are able to predict what a line should look like, it will help them to recognise when they have made a mistake. If they understand this concept fully now, then they will find it easier when calculating gradients in the future.

What do you think? 4.4A

1 requires students to explain why this line cannot have a negative gradient. They could consider why somebody might think that it does: as the graph intersects the *y*-axis below 0, students can sometimes confuse this with meaning that the graph has a negative gradient.

2 Students identify a line with a negative gradient from multiple representations. Graphs and equations have featured prominently throughout this chapter, and students should be gaining confidence at linking the two. They should now consider how the gradient can be predicted from a table of values. They should recognise that if *y* decreases as *x* increases, it means that the line has a negative gradient.

Consolidate – do you need more?

These questions revisit the key points covered in the previous exercise. Students use a function machine to support them in working out the coordinates of points on a line. They should be encouraged to explore how one question supports the next. If further practice is needed, students could be given the equations of more lines and asked to repeat these steps to plot them.

Stretch – can you deepen your learning?

1 looks at whether a line has a positive or a negative gradient and then considers the differences between lines with positive and negative gradients. If necessary, students could draw the lines $y = 2x$ and $y = -2x$ accurately before this question for support.

2 Watch out for students thinking that a line has a negative gradient only if it goes through the second and fourth quadrants. They could be encouraged to draw a graph that appears to support this statement, as well as one that disproves it.

3 This links to the idea covered in the "What do you think?" section. Students look at how a gradient can be seen from a table of values. They should recognise that the *x* and *y* values are both decreasing from left to right, so the line has a positive gradient. They could draw the line to verify this if necessary.

4 Students use their learning from the block so far to identify which equation matches each line. They should be encouraged to explain each step in their thinking as they work, and to say which one they chose to start with and why they chose it. Watch out for students struggling with C and H as these lines have not featured much in this chapter. Once they have correctly matched the lines to the equations, students could be stretched to consider "What are the coordinates of the point of intersection of line … and line …? How do you know?" For example, lines C and H intersect at (–5, 4) because $x = -5$ and $y = 4$.

Reflect

Students should explain their learning in their own words as this will clarify their thinking. They should be able to sketch graphs to support their reasoning.

4.5 Lines of the form $y = mx + c$

Small steps

- Plot graphs of the form $y = mx + c$
- Link graphs to linear sequences

Rationale

Here students are introduced to lines of the form $y = mx + c$ for the first time. The focus here is on completing a table of values and drawing the lines: students are not required to interpret the gradient and y-intercept from the equation at this stage, as this will be covered formally later in the curriculum. Students could be encouraged to explain how their graph is linked to the equation, e.g. positive/negative gradient or crossing the y-axis above/below 0 using what they have covered in previous chapters, but they should not go beyond this. Students will use their understanding of the relationship between the x and y values of any point on a line, and correct mathematical language is encouraged to support this. For example, reading "$y = 2x + 1$" as "y is equal to 2 times x plus 1" supports understanding of why, at every point on this line, the y value is 1 more than twice the x value. Students then look at graphs of linear sequences and should be able to explain the similarities and differences. It is important that they understand that the corresponding graph would go through the points generated by the sequence, but that the sequence itself is just a series of points, not a straight line, as there are no interim values.

Things to look out for

- Writing coordinates the wrong way round
- Not paying attention to the scale used on an axis
- Getting the x and y values confused in a pair of coordinates
- Incorrect calculations with directed number
- Not extending a line to the full width of the grid, but instead only drawing it between the points plotted
- Not using the correct order of operations, e.g. evaluating $2(x + 1)$ rather than $2x + 1$
- Drawing a line through the graph of a sequence

Interleaving opportunities

- Working with coordinates in all four quadrants
- Lines parallel to the axes
- Substitution
- Algebraic notation
- Solving equations
- Directed number
- Sequences

Models and representations

- Cubes or counters are useful for students to create sequences practically. They can link this to graphs in the form $y = mx + c$.

Behind the questions

Are you ready?

These four questions cover the essential prerequisite knowledge for this lesson. Questions 1 and 2 look at substitution into a two-step expression. Question 1 uses a function machine to support understanding, whilst question 2 is more abstract; if necessary, students could draw function machines to support their working in question 2. Questions 3 and 4 look at linear and non-linear sequences, as these will be encountered in Practice 4.5B, when students link graphs to linear sequences.

Practice 4.5A

1 Students combine what they have covered so far in this block to look at lines of the form $y = mx + c$. Reading the equation as "y is equal to 5 lots of x plus 3" or "y is equal to 5 times x plus 3" can support understanding.

2 Students complete tables of values and plot lines of the form $y = mx + c$. The lines in this question are the same as those in part d of question 1. Students have already explained the relationships between the x and y values of the points on these lines, and this will help them to find the coordinates of points on the lines.

3 Students now look at four different lines that each have a gradient of 3. Similarly to when they were working with lines of the form $y = x + a$, students should notice that the lines are parallel and cross the y-axis at different points. They could be encouraged to consider which part of the equation must account for the similarities and which part is responsible for the differences, but they do not need to formalise this.

4 uses a similar idea to question 3 but this time with lines that have the same *y*-intercept. Students should notice that all of the lines meet at exactly one point, which has coordinates (0, 4). Again, they could be encouraged to consider which part of the equation is responsible for this.

5 Students practise plotting more graphs of the form $y = mx + c$ and then finding points of intersection. Although the points can be read off from the graph, students could "show that" their points are correct by substitution.

What do you think? 4.5A

1 asks students to explain a mistake that has been made and correct it. Samira has made a mistake when calculating with directed numbers. As this is a really common error, it is important that students can recognise when a line is drawn incorrectly to avoid them making the same mistake.

2 looks at this concept from a slightly different angle. By this stage in the chapter, students will have completed plenty of tables of values and then used them to draw lines, but it is also important that they can read coordinates from a graph. Here they need to complete a table of values when given a line.

Practice 4.5B

1 Students should be encouraged to spot connections between the parts of this question. In part b, they explain that the sequence is increasing by 1 each time, and they should be able to spot this in their table of values in part c. It is important in part d that students only plot the points and do not draw a line through them. They should understand why this is: in any sequence, the position is always an integer and so drawing a line would not make sense. However, they could use their graphs to state how many squares are in the 8th term, and could be encouraged to draw the term to verify their answer.

2 Students should notice that their table of values is similar to that in question 1, except for the row headings.

3 Students compare the similarities and differences between questions 1 and 2. They should recognise that the line drawn in question 2 ($y = x + 2$) goes through the points generated by the sequence in question 1.

4 to **6** build on the ideas from questions 1 to 3 but start to link the equations of the lines to the diagrammatic representation of each sequence. Students should recognise that the coefficient of x is how many are added each time, and the "a" from "$+ a$" is the first term of the sequence.

What do you think? 4.5B

These three questions encourage students to think about the connection between graphs and linear sequences more deeply. In question 1, they are required to find the equation of the line that goes through the points generated by the sequence, whilst in question 2 they think about what the sequence could be given the line that goes through the points. Question 3 further highlights why a straight line does not represent a sequence, but instead just goes through the points generated by one.

Consolidate – do you need more?

These questions revisit the key points covered in the previous exercises. Students use a function machine to help them to work out the coordinates of points on a line. If further practice is needed, students could be given the equations of more lines and encouraged to repeat these steps to plot them.

Stretch – can you deepen your learning?

1 Students use all the learning so far in this chapter to match each graph to its equation. They should be encouraged to explain each step in their thought process.

2 Students should recognise that the line only briefly passes through the fourth quadrant and so this point is not on the line. Substitution could be used to confirm this.

3 to **5** link back to previous learning on forming and solving equations and ensure that students understand the link between coordinates and an equation.

Reflect

Students should explain their learning in their own words as this will clarify their thinking.

4.6 Non-linear graphs (H)

Small steps

- Explore non-linear graphs (H)

Rationale

Here students are introduced to non-linear graphs for the first time. Understanding why non-linear graphs do not form straight lines can be supported by looking at non-linear sequences, i.e. sequences which do not increase/decrease by the same amount each time. This chapter focuses specifically on quadratic and cubic graphs; students are required to work out coordinates and draw these types of graphs. They should join the points using smooth curves. Other non-linear graphs can be explored using computer graphing software. Correct mathematical language should be used to support understanding of the relationship between the *x* and *y* values at any point on these graphs. For example, reading "$y = x^2$" as "*y* is equal to *x* squared" can support understanding of why, at every point on this graph, the *y* value is the square of the *x* value.

Things to look out for	**Interleaving opportunities**
• Writing coordinates the wrong way round • Not using the correct order of operations, e.g. working out $(2x)^2$ rather than $2x^2$ • Using a ruler to connect the points on a non-linear graph • Incorrect calculations with directed number	• Working with coordinates in all four quadrants • Substitution • Algebraic notation • Solving equations • Directed number • Sequences

Models and representations

- When working with graphs in this form students may need reminding of the order of operations to avoid calculation errors.

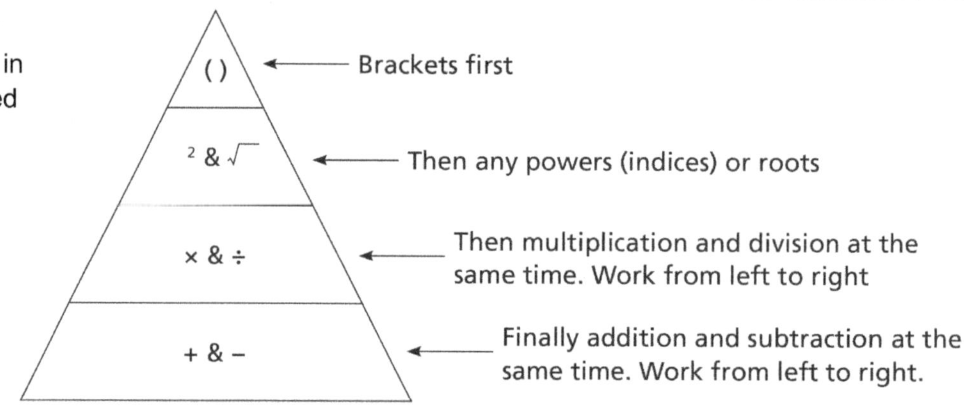

Behind the questions

Are you ready?

These four questions cover the essential prerequisite knowledge for this chapter. Questions 1 and 2 look at the difference between "linear" and "non-linear" in the context of sequences. As well as being able to recognise which is which, students need to be able to continue both types of sequences so that they can see how they both change. Question 3 looks at squaring and cubing numbers, including directed numbers, and question 4 looks at substitution in more complex expressions. In part e of question 4, watch out for students giving 324 rather than 108 because they have used the incorrect order of operations.

Practice 4.6A

1 recaps learning on the line $y = x$ from earlier in the block. Students should be able to plot the graph but also to explain the features that make it linear. They should know that it is linear because it forms a straight line.

2 Students are now introduced to the graph of $y = x^2$. Watch out for incorrect calculation with directed numbers. In part c, students are told that the graph isn't linear but they need to explain that they can tell this because it is not a straight line.

3 This is similar to question 2 but now students look at the graph of $y = x^3$. Again, watch out for calculation errors, particularly when working with directed numbers. In part c, students need to decide whether the graph is linear or non-linear and should be able to explain fully why it isn't. Looking at the sequence formed by the y values can support reasoning, if necessary, by linking back to the idea of non-linear sequences covered in the "Are you ready?" section.

4 and **5** allow students to look at different variations of the graph $y = x^2$. They should be encouraged to reflect on the similarities and differences between the graphs and to compare them with $y = x^2$, which was drawn in question 2. They should realise that the similarities arise because all of the equations are in terms of x^2 and that the differences are a result of the operations applied to x^2. In question 5, students might make comparisons with lines of the form $y = x + a$: the value of a is where the graphs cross the y-axis, and the same relationship is true here.

6 gives students the opportunity to look at different variations of the graph $y = x^3$. They should be encouraged to reflect on the similarities and differences between the graphs and compare them with $y = x^3$, which was drawn in question 3. They should notice that the similarities arise because all of the equations are in terms of x^3 and that the differences are a result of the operations applied to x^3.

What do you think? 4.6A

1 allows students to explore some common mistakes. Rhys has found the square of a negative number to be negative and then joined his points using a ruler rather than using a smooth curve.

2 and **3** Students have now been exposed to several non-linear graphs and should now be able to identify when a graph is non-linear and when it is linear, both from the equation and from a sketch of the graph. Watch out for students thinking that $y = \frac{15}{4x}$ is linear because there is no power on the x. Graphing software can be used to support this, or showing that the sequence formed by the y values is non-linear. Also, in question 2, watch out for students saying that $y = 3x^2 + 4x - 2x^2 - x^2$ is non-linear because it contains x^2 terms rather than first simplifying and realising that these terms cancel.

Consolidate – do you need more?

These questions revisit the key points covered in the previous exercise. Students use a function machine to help them to work out the coordinates of points on a graph. They should be encouraged to explore how one question supports the next. If further practice is needed, students could be given the equations of more non-linear graphs and asked to repeat these steps to plot them.

Stretch – can you deepen your learning?

1 Students use their knowledge of quadratic graphs to match each graph to its equation. They should be able to explain their reasoning and the order in which they matched them. If necessary, students should look back at their working for questions 4 and 5 of Practice 4.6A to support them.

2 is similar to question 1 but looks at cubic graphs. Again, students could look back to question 6 from Practice 4.6A to support their thinking.

3 encourages students to make the connection between the equation of a graph and the relationship between the coordinates of any point on the graph in order to find a missing value.

Reflect

Students should explain their learning in their own words as this will clarify their thinking. They should be able to sketch graphs to support their reasoning.

4.7 Midpoint of a line segment (H)

Small steps

- Find the midpoint of a line segment (H)

Rationale

Here students are introduced to finding the midpoint of a line segment. They should first consider finding the midpoint on a number line to find the most efficient method before transferring this to working in two dimensions.

Things to look out for	**Interleaving opportunities**
• Calculation errors • Finding the midpoint of the *x* and *y* values in one pair of coordinates rather than using the *x* values of two pairs • Directed number errors • Finding half of the difference but not adding it on • When given an end point and the midpoint, finding the midpoint of these two rather than finding the other end point	• Four operations • Directed number • Number lines • Median and mean • Fractions • Decimals • Algebraic notation • Ratio • Shape

Models and representations

- A number line can be used to support understanding of finding the midpoint of two numbers, a skill which can then be transferred to finding the midpoint of a line segment.
- For example, to find the midpoint of 12 and 37:
  The difference between 12 and 37 is 37 – 12 = 25
  This means that the difference between both 12 and 37 and their midpoint must be half of this, which is 12.5.
  So the midpoint of 12 and 37 is 24.5 (12 + 12.5 = 24.5 and 37 – 12.5 = 24.5)

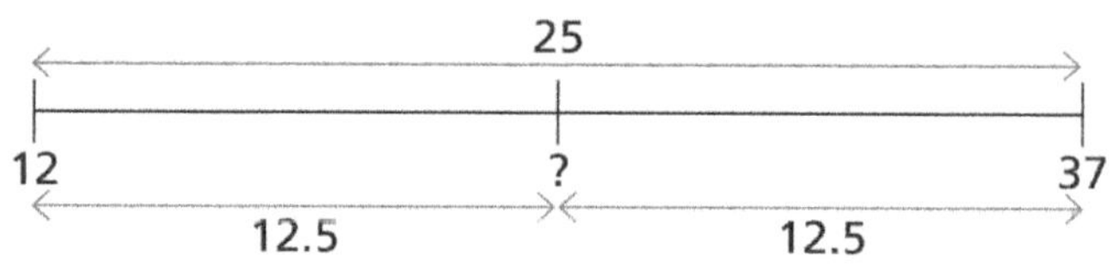

Behind the questions

Are you ready?

These three questions cover the essential prerequisite knowledge for the chapter. In question 1 students find the midpoint on a number line. In parts a, b and c, students could draw the other values on the number line for support, but they should realise by part d that this method won't always be efficient. In the first four parts, the starting point is 16, and students should be encouraged to reflect on the similarities and differences between the question parts. By the time they get to parts d to h, students should be selecting the most efficient method for each one. They could be encouraged to work out the answers in different ways, and you could discuss which they think are best. Watch out for students not giving decimal answers where these would be correct, because they think that coordinates can only be integers. Question 2 looks in a more abstract way at finding the middle value of two numbers, but students could sketch a number line for support if they need to. In question 3, students are given the answer but they need to use a variety of methods to show it; they should reflect on the pros and cons of each method.

Practice 4.7A

1 Students find the midpoints of horizontal and vertical line segments. As the lines are drawn, they could find the coordinates of each midpoint by looking at the coordinate grid and counting squares, but they should be encouraged to find a method for calculating the coordinates. Students should recognise that for a horizontal/

vertical line segment, the *y*/*x* value of the midpoint is the same as at the end points and reflect on why this is the case.

2 builds on question 1 but this time the line segments aren't drawn. The numbers are deliberately kept quite small so that students can choose to sketch these to check their answers if they want to. Again, the line segments are all horizontal or vertical, and students should be able to explain how they can tell from the coordinates of the end points that the line segment is horizontal/vertical and then be able to explain what this means about the coordinates of the midpoint.

3 and **4** take a similar approach to questions 1 and 2 but this time look at diagonal line segments. This is the first time that students have been required to find the midpoint of both the *x* values *and* the *y* values. Use of a coordinate grid in question 4 can provide support if needed.

5 requires students to work abstractly to find the midpoint of a line segment. Notice that all the pairs of numbers for which students are required to find the midpoint have already been used in the "Are you ready?" section. Drawing students' attention to this can help to reduce the cognitive load. Instead, they can focus on which coordinates correspond rather than on the actual calculation.

6 Watch out for students adding all of the *x* values and halving, rather than correctly identifying the corresponding coordinates to find the midpoint of each line segment.

What do you think? 4.7A

1 Students find the end point given the midpoint. Watch out for students finding the midpoint between A and each point given in the question. Parts a to c deliberately use small numbers so that a diagram can be used to support understanding if needed. By the time students get to parts d and e, they need to have developed a reliable abstract method that they can explain.

2 Students could be encouraged to generalise the relationship between the *x* and *y* values at each point using the fact that the *x* values of X and Y will be equidistant from 17 and the *y* values of X and Y will be equidistant from –21. They should recognise that there are an infinite number of pairs of coordinates.

Consolidate – do you need more?

These questions revisit the key points covered in the previous exercise. Students could use a grid to help them to find the coordinates of each midpoint. Ensure that they write down the coordinates of the midpoint and don't just mark this with an X on their diagrams. Once they have completed each question part, they could consider the relationship between the end points and the midpoints and try to work out the midpoints abstractly.

Stretch – can you deepen your learning?

1 and **2** interleave prior learning on fractions, decimals and algebra with finding the midpoint. Students should be able to explain each step of their thinking.

3 encourages students to deduce a general rule for finding the midpoint of any pair of coordinates. They could test their answers for some given values of *p*, *q*, *w* and *z*.

5 Students could draw a diagram to support their thinking in this question. Watch out for students "counting diagonally" to find an incorrect side length for the outer square in part b. Instead, they should find the area of the four right-angled triangles that border the smaller square and use this to help them find the area of the larger square.

Reflect

Students should explain their learning in their own words as this will clarify their thinking. They should be able to give examples and draw diagrams to support their reasoning.

5.1 Scatter graphs

Small steps

- Draw and interpret scatter graphs
- Understand and describe linear correlation
- Draw and use lines of best fit
- Identify non-linear relationships

Rationale

The purpose of this chapter is to introduce students to scatter graphs. This is the first time they will have experienced this type of graph. Previously, they have looked at bar charts, pie charts and line graphs. Examples of pairs of variables that are appropriate/not appropriate to represent using a scatter graph should be discussed during this chapter.

Things to look out for

- Plotting points incorrectly
- Drawing or reading scales incorrectly
- Misunderstanding the language used, e.g. "linear" and "correlation"
- Joining the line of best fit to the origin or drawing it like the line $y = x$ rather than through the given points
- Getting mixed up between describing the relationship and stating the correlation

Interleaving opportunities

- Work with coordinates
- Number lines, scales and place value
- Accurate drawing
- Linear graphs and sequences
- Estimation

Models and representations

- Scatter graphs are the only representation used in this chapter. Students need to be able to recognise those showing positive, negative and no correlation.

Behind the questions

Are you ready?

These two questions provide an opportunity for students to practise plotting and reading coordinates.

Practice 5.1A

1 is designed to give students practice at plotting points on a scatter graph. Notice that most of the coordinates are even so that they are straightforward to plot. You could provide a pre-printed copy of the grid to save time; however, it is good practice for students to learn how to scale a graph independently.

2 focuses on interpreting a scatter graph rather than plotting the points. This is the first time students have experienced a negative relationship. Encourage students to discuss their experiences of this in real life.

3 Students may find this question difficult as they need to think about what scale their graphs will use. Part a i is intentionally similar to the graph used in question 1 so that they can use the graph there as a reference point. Part a ii would work well with a scale going up in 5s starting at zero, whereas, in part a iii, for the first time the graph does not start at zero. Highlight the use of a broken scale here.

4 exposes students to scatter graphs with no correlation. At this point there is no need to introduce the word "correlation"; instead, just focus on the fact that there is no relationship between the two variables. Encourage a class discussion about other variables that would show no clear relationship.

What do you think? 5.1A

1 gives students the chance to think about data that might interest them.

2 introduces students to the idea of outliers and allows them to think about why sometimes there may be exceptions to a rule.

Practice 5.1B

1 and **2** introduce the idea of correlation. Students should be able to recognise correlation and think of relationships that would produce positive, negative or no correlation.

3 looks at common misconceptions for drawing a line of best fit. Looking at these before they draw their own should help prevent students from making those mistakes.

4 This is the first time that students need to draw their own line of best fit. It may be worth providing them with a copy of the graph in this instance. It is also the first time that they have been asked to find an estimate from a graph. Students may get slightly different answers.

5 Part c is an interesting point for students to consider and shows that scatter graphs can have restrictions. Students should notice that if the trend continues, then an office space would end up having £0 rent, which is unrealistic. In addition, the further away from one city you travel, the more likely you are to end up closer to a different city, and therefore the rent would begin to increase again.

What do you think? 5.1B

1 Encourage students to work backwards to create a table of values from the scatter graph. This will make the data easier to interpret. Students studied the mean, median and range in Book 1, but they may need a reminder.

2 You may need to explain the meaning of the term "non-linear" in this context.

Consolidate – do you need more?

This provides further practice at drawing scatter graphs, interpreting scatter graphs and recognising correlation. There is also an opportunity in question 2 to discuss the idea of a correlation being strong or weak.

Stretch – can you deepen your learning?

1 Students could approach this question in a variety of ways. They could draw the graph and extend the axis to include values up to 2025. They could then use a line of best fit to estimate the price of the magazine in 2025. Alternatively, they could find the price increase between 2017 and 2021 and assume that the price will increase by the same amount in the next four years. Encourage students to reflect on which method would be more efficient.

2 In part d, it is important for students to recognise that there was no data collected in this range and therefore the graph wouldn't be helpful. Students may have chosen to start their x-axis at 70, and therefore this point wouldn't even fit on the graph.

Reflect

Students should recognise the pros and cons of using a scatter graph to represent data. They are useful for showing trends and relationships and allowing you to make predictions, but using them for sets of data with no relationship is not helpful. However, sometimes scatter graphs can help to establish the fact that there is no relationship between two variables.

5.2 Types of data

Small steps

- Identify different types of data
- Read and interpret ungrouped frequency tables

Rationale

In this chapter, students will be introduced to some new language relating to data. Before this point, they may not have realised that there are different types of data that can be categorised in a variety of ways. Once students are confident with the different types of data, they will begin to organise discrete data into frequency tables. Students should be encouraged to recognise the benefits of recording data in a frequency table, rather than just as a list, particularly when working with large quantities of data. They should also understand the limitations of ungrouped frequency tables and that they are not appropriate for continuous data.

Things to look out for

- Thinking that money is continuous because it is a decimal
- Thinking that shoe size is continuous because it is linked to the length of your foot
- Getting "qualitative" and "quantitative" mixed up or not being able to say the words aloud
- Adding up the frequency column without considering what it represents, for example the number of goals scored

Interleaving opportunities

- Addition and subtraction
- Tally charts
- Multiplication
- Units of measure
- Averages
- Different types of charts and graphs

Models and representations

- A frequency table is a way of organising data so that it is easier to read and interpret.

Shoe size	Frequency
4	12
5	14
6	7

Behind the questions

Are you ready?

These questions remind students how to interpret and set up tally charts. They should be familiar with this from primary school. Use these questions to check their understanding of the word "frequency".

Practice 5.2A

1 to **3** focus on the difference between discrete and continuous data. Students might be tempted to say that the answer to part e of question 1 is "discrete" because they're likely to have only ever drawn and measured angles that are a whole number of degrees. This is a good opportunity to discuss how rounding is often used when working with angles.

4 aims to give students some ideas to help them to remember the new vocabulary. A good discussion point could be around numerals being written in words, such as a pet being "two years old". Does this make the data qualitative because it is written in words?

5 requires students to interpret data presented in different types of charts, and gets them to consider which chart is appropriate for each type of data. In part b, some students might think that this is qualitative data because countries are represented on the *y*-axis, but they should be encouraged to notice that the data is about the number of medals for each country, and that this is actually discrete quantitative data.

6 Part c gets students to think about the limitations of ungrouped frequency tables. Although Darius could have had many more rows to try to cover all the possible colours of cars, this wouldn't be a very efficient method. Students need to think of ways of ensuring that the data collection process is as efficient as possible.

7 gets students to record data in a table. Watch out for students missing pieces of data and encourage them to take a systematic approach to avoid making silly mistakes.

8 highlights a common misconception. Students need to think carefully about what the question is asking. They are asked for the total number of passengers, so should recognise that if 21 cars each contain 2 passengers, then this represents a total of 42 passengers. Using contexts that are familiar to students will help them to grasp this concept.

9 Building on question 8, this question will help you to check that students have appreciated the common misconception they have just discussed. Students may find the row for 0 pens confusing as this row has a frequency but it won't be included in their answer because they are only interested in the number of pens, not the number of students. This brings in some prior learning.

What do you think? 5.2A

2 gives students the opportunity to think about and explore what they can and cannot interpret from a frequency table. Although they know how many words have 4 letters, they don't actually know what any of the words are. It is important that students understand that there is a limit to what they can find out from the table. This will be built on further when looking at grouped frequency tables.

Consolidate – do you need more?

This provides further practice to help students consolidate their understanding. The contexts chosen are intended to be simple, to ensure that students are familiar with them and can use them to support their learning.

Stretch – can you deepen your learning?

These questions require students to reason and solve problems about data and tables. Question 2 looks difficult at first glance as two pieces of information are missing. If students have a thorough understanding of zero in this context, they should notice that whilst the two missing pieces of data must sum to 9, the remaining 12 goals must all be in the row for 2 goals.

Reflect

This further reinforces the sticking point for this topic. It will ensure that students understand the difference between when they multiply to find the total and when they simply need to add up the frequency column.

5.3 Grouping data

Small steps

- Read and interpret grouped frequency tables
- Represent continuous data grouped into equal classes

Rationale

In this chapter, students will work with discrete and continuous data that needs to be grouped. It is important that students see why, sometimes, discrete data should be grouped. When grouping continuous data, students are expected to use the inequality symbols. They should understand why these do not need to be used for discrete data. Ensure that students can read statements aloud using the symbols. Often students cannot do this, which leads to a general lack of understanding of the topic and what a grouped frequency table actually represents.

Things to look out for	Interleaving opportunities
• Not understanding the meaning of the inequality symbols • Duplicating options between rows in a table • Not understanding when data needs to be grouped and when it does not • Not choosing an appropriate number of groups and not realising that there may be several correct ways of grouping data	• Averages • Units of measure • Use of the inequality symbols • Accurate drawing • Fractions and percentages

Models and representations

- Grouped frequency tables can contain discrete or continuous data.

Grouped discrete data

House number	Number of houses
0 to 9	3
10 to 19	8
20 to 29	8
30 to 39	9

Grouped continuous data

Time taken, t (hours)	Frequency
$2 < t \leqslant 4$	15
$4 < t \leqslant 6$	26
$6 < t \leqslant 8$	38
$8 < t \leqslant 10$	40

Behind the questions

Are you ready?

These questions revisit the difference between discrete and continuous data, the range and the inequality symbols. Ensure that these concepts are secure before moving into the main part of the chapter. In particular, students may find it difficult to read and interpret statements written with the inequality symbols. Ensure that you have some additional questions like this prepared in case students need a lot of support. Devoting time to this now will speed up their understanding later in the chapter.

Practice 5.3A

1 is a basic but important step which is often missed. Ensure that students understand the difference between a grouped frequency table and an ungrouped frequency table.

2 Here students begin to think about why data may need to be grouped. You could start by asking students to put the data into tables. They should soon realise that some sets of data will produce numerous rows and therefore should be grouped. Try not to let students dwell on creating a context for each set of data.

3 The table is partially complete to allow students to begin to design their own grouped frequency tables. Discuss why the first group starts at 700 and not zero. Where is a sensible place for the final group to end? Should the final group be 900 to 985 or 900 to 999? Talk to students about the importance of groups having an equal range: therefore, it is more sensible for the final group to be 900 to 999.

4 The table in this question is presented differently to ensure that students are exposed to all possibilities. Parts a and b are simple interpretation whereas parts c and d will need some extra thought from the students. In part c, students may be tempted to try to include part of the 40–59 group if they do not fully understand the word "fewer". They need to know that "fewer than 40" means less than 40 and therefore would end at 39. Part d helps students to recognise the limitations of a grouped table: some information is impossible to find.

What do you think? 5.3A

This question requires students to create their own grouped frequency table. Allow students to struggle, experiment and make some mistakes rather than guide them through it. When they have completed the task, compare tables and strategies. Would they go about it differently next time? Why? Why not? Is there one perfect answer or are several tables correct and appropriate? Students should recognise the pros and cons of having fewer groups with a larger range compared with more groups with a narrower range.

Practice 5.3B

1 reinforces the meaning of the inequality symbols and ensures that students can interpret these before getting started.

2 emphasises the need for using inequality symbols for continuous data. Question 1 should have nicely led students into thinking about setting up groups using the symbols, and they should recognise that it is impossible to tell where in Junaid's table someone with a height of 145 cm would be represented.

3 introduces the language "class interval". Look out for students writing down the frequency rather than the class interval here. Part d provides the opportunity to revisit fractions. The answer will simplify, so encourage students to compare their answers and discuss why they may be different.

4 Here students need to set up their own groups. Watch out for students writing the symbols the wrong way around. If this happens, encourage them to say aloud what they have written. Does it make sense? There is also an opportunity to revisit percentages in part c. The total frequency is intentionally 20 to make it straightforward for students to work out the percentage with or without a calculator.

What do you think? 5.3B

These two questions further emphasise the meaning of a grouped frequency table and some of the limitations when trying to extract data from one. When estimating in question 1, talk about what students' estimate must be to help them get started. For example, their estimate must be less than 163 and greater than 78. Why is this the case? How does this help them to work out an estimate?

Consolidate – do you need more?

This provides extra practice on the different types of grouped frequency tables. The questions highlight the key misconceptions that students may have with this topic.

Stretch – can you deepen your learning?

1 provides a modern context for students to interpret. A common mistake would be for students to complete the table using the information given rather than considering that the information overlaps the class intervals. Allow students to make this mistake and then discuss why it is incorrect. They can self-check their answers by checking that the frequencies add up to 80.

Reflect

This brings together students' learning on types of data and displaying it in an appropriate grouped frequency table.

5.4 Two-way tables

Small steps
• Represent data in two-way tables

Rationale

In this chapter, students will extend their learning of frequency tables to two-way tables. A secure understanding of addition and subtraction is vital for this topic. Students may struggle, as there is not a clear process to follow when completing a two-way table. They will have to problem-solve and make decisions depending on the information they are given.

Things to look out for

- Calculation errors
- Working out any missing values in a row/column by just finding the total of the other values in the same row/column rather than considering its position and what it represents
- Not recognising when the total is given in the question rather than in the table
- Working out the "total total" to be double its actual value by finding the total of the entire bottom row *and* the entire last column

Interleaving opportunities

- Addition and subtraction
- Fractions of an amount
- Percentages

Models and representations

- Two-way tables present data from two categories. There can be numerous options within each category, and students should be exposed to a range of these.
- This two-way table has two categories with two options each.

	Owns a pet	Does not own a pet	Total
Year 7			
Year 8			
Total			

- This two-way table has two categories with three options for type of sport and two options for gender.

	Football	Hockey	Netball	Total
Boys	7		6	24
Girls		3		
Total	12			40

Behind the questions

Are you ready?

Questions 1 and 2 check that students are confident with basic addition and subtraction, as they will need this throughout the chapter. If students are not confident, consider allowing them to use a calculator. This will ensure that they don't overly worry about the calculations and instead focus on the new learning of two-way tables. Question 3 ensures that students can interpret a basic frequency table before moving on to more complex two-way tables.

Practice 5.4A

1 provides the opportunity for students to demonstrate understanding of what a two-way table is and what it represents. The focus is not on calculating values, but on reading values from a completed two-way table. Students could be stretched to consider things such as "If one more purple parallelogram is added, which cells in the two-way table would be affected?"

2 Here students are required to complete a two-way table for the first time. The information in the question is given in the order that it will be entered into the table. Students are only required to work out the totals. They should be encouraged to explain how they know that they only need to perform addition in this case.

3 provides students with practice at completing two-way tables. This is the first time that they will be independently required to perform both addition and subtraction calculations to complete the two-way tables. They should be encouraged to explain how they know what calculation they need to perform in each case. If basic number skills are a barrier, calculators could be used to ensure that the focus is on completing and interpreting the two-way tables.

4 In this question, students might think at first glance that they do not have enough information to complete the two-way table. The total is given in the opening line of the question and some students often miss this. If this happens, they should be encouraged to reread this sentence and think about what the given value represents.

5 requires students to draw out their own two-way table for the first time. Students often struggle to identify what categories are needed in the table. If this issue arises, they should be encouraged to think carefully about what information they've been given to be able to interpret the categories needed. Watch out for students drawing their tables in different orientations. This can lead to some students thinking that they have made a mistake because their answer doesn't look exactly the same as other people's. However, reading out several values can help students to realise that the orientation doesn't affect the answers.

What do you think? 5.4A

1 requires students to think about what you can and can't find out from a two-way table. The statements are designed to highlight the limitations of two-way tables.

2 Here students might recognise that this two-way table looks different from the ones they've seen previously as it doesn't have a totals row and totals column. This question is designed to ensure that students look at the whole picture and realise that although you can just read off numbers from a two-way table, it is actually more useful to look at the bigger picture and interpret proportions.

Consolidate – do you need more?

These two questions provide extra practice for students who need it. Question 1 supports students by giving them the shapes so that they can clearly see where each number in the two-way table has come from. Students then transfer this knowledge to question 2 to help them to complete the two-way table.

Stretch – can you deepen your learning?

1 incorporates fractions of an amount and ratio language into two-way tables. Students need to carefully consider the information given and should be encouraged to check all their values once their two-way tables are complete.

2 Watch out for students thinking that the totals in the two tables should be the same. They should notice that whilst the first table shows the total number of scoops sold, the second table shows the total number of ice creams sold (some of which have multiple scoops).

Reflect

These questions encourage students to think about how they can check their own answers and what strategies they may need to use to complete a two-way table.

6.1 Sample spaces and probability

Small steps

- Construct sample spaces for 1 or more events
- Find probabilities from a sample space

Rationale

The purpose of this chapter is to build on students' knowledge of probability from Book 1 and to start to consider sample spaces as a way of recording outcomes and calculating probabilities. Students should recognise the benefit of using a sample space diagram as a way of working systematically and limiting the risk of missing out any outcomes. Once students can record possible outcomes using sample space diagrams, they can then start to use them to calculate probabilities. They are introduced to the notation "P(event)" as a way of asking "What is the probability of this event happening?" and should be encouraged to read it as such rather than just "P brackets event close brackets", so that they fully understand the meaning of the notation.

Things to look out for

- Students not working systematically and hence not including all the possible outcomes in their sample space
- Including the top row and first column in their count when working out the total number of possible outcomes from a sample space diagram (particularly when they are fully numerical)
- Using the number of times an outcome does not happen as the denominator of a fraction when calculating probability rather than using the total number of outcomes
- Misinterpreting questions such as P(greater than 4) and including outcomes that are equal to 4

Interleaving opportunities

- Basic probability
- Types of numbers (e.g. odd, even, prime, etc.)
- Money
- Fractions
- Decimals
- Percentages
- Two-way tables

Models and representations

- A sample space diagram is used for support throughout. It encourages students to work systematically when recording all possible outcomes of two events. This sample space diagram could be used to record the outcomes of these two spinners.

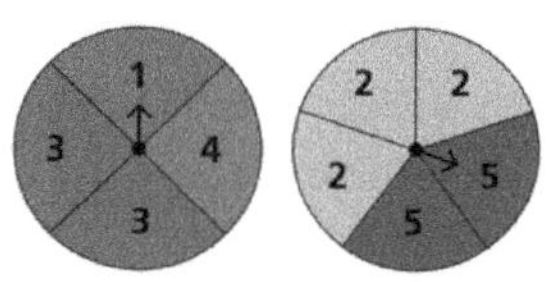

		Spinner 2				
		2	2	2	5	5
Spinner 1	1					
	2					
	3					
	4					

Behind the questions

Are you ready?

These four questions provide an opportunity for students to practise the key skills needed in this chapter. They covered probability in Book 1 and these questions will remind them of this. Question 1 looks at basic probability, and students should be able to verbalise why it is important that the dice used here is fair. In question 3, students are reminded of the crucial point that probabilities sum to 1.

Practice 6.1A

1 looks at two different ways of completing a sample space diagram. In part a, students are required to just list the outcomes and in part b they are required to find the total value of the two coins. Students could be asked to consider what is the same and what is different about these before moving on to part c. They may realise that the outcomes in the table in part a provide enough information to work out the corresponding outcomes in part b, whereas just the outcomes from the table in part b would not be enough information to work out the corresponding outcomes in part a. In part c, students will have to choose between the two tables when calculating the probabilities and they will need to reflect on these choices in part d.

2 provides the opportunity for students to see common misconceptions. They often think that when you can either win or lose a game, the probability of each must be one half. This question reinforces the idea that games and spinners aren't always fair, and the weighting of the sections in this case affects the probabilities. When completing part b, students could reflect on this misconception from part a to consider why it is important that "win" and "lose" are recorded multiple times for spinner B.

3 Here students consider the sample space diagram for rolling two dice and adding their scores. In part d, students will need to calculate the probability of each event before comparing their answers. This provides the opportunity to recap some basic fraction knowledge.

4 Here students are required to draw their own sample space diagram for the first time and then calculate probabilities from it. Students might draw their sample space diagrams in different orientations, and it's important to discuss why this does not make a difference. Comparing their probabilities with a partner who has drawn theirs differently can support this understanding.

5 highlights the idea that if a card is not replaced, then this affects the possible outcomes for the second card. This can be conceptually quite challenging for students, and to support this question they could be encouraged to work with a partner and five number cards so that they realise why some of these outcomes are not possible. They should realise that the values in the diagonal from top left to bottom right are not possible, as the players cannot both take the same number if there is only one of each number in the set of cards. Students could be stretched to consider how Rhys and Amina would have to change their process for this sample space diagram to be correct.

What do you think? 6.1A

These three questions get students to look more deeply at sample space diagrams and probabilities. In question 1, students should consider whether it is beneficial to draw a sample space diagram or if it is more efficient to just record the sample space when there is only one event. Although part b is better suited to a sample space diagram, the outcomes could still be listed, but for part c students should be able to see the benefit of drawing a sample space diagram. Question 3 highlights some of the common misconceptions surrounding sample space diagrams and probability, and gets students to explore them. The use of number cards could support students so that they can try the activity out before commenting on these statements.

Consolidate – do you need more?

These questions revisit the key points covered in the previous exercise. Question 1 gives the sample space diagram so that students only need to focus on the probabilities, whilst question 2 requires them to draw their own sample space diagram. If further practice is needed, students could be challenged to work with a partner to come up with their own sample space diagram for different events and calculate probabilities from it.

Stretch – can you deepen your learning?

1 looks at sample space diagrams in a different way. This links back to previous learning on two-way tables. Students should recognise that although they could complete the cells as (0, 1), etc., they wouldn't be able to calculate the probabilities from this alone because the numbers of people in each category are important.

2 Here students could be encouraged to draw a sample space diagram and then think about how many of the outcomes need to be greater than 10 and how many need to be odd, and use this to help them fill in the values.

3 requires students to recognise that when adding a third event, they will need to draw two sample space diagrams: one for the outcomes of the first two events and a second one to allow them to record the outcomes of the third event. For each outcome of the first two events, the third outcome could either be heads or tails, so there will be twice as many outcomes when flipping three coins as with two coins.

Reflect

Students should explain their learning in their own words as this will clarify their thinking. They should be able to identify the benefits of using a sample space diagram for recording outcomes and calculating probabilities.

6.2 Using diagrams to find probabilities

Small steps

- Find probabilities from two-way tables
- Find probabilities from Venn diagrams
- Use the product rule for finding the total number of possible outcomes (H)

Rationale

The purpose of this chapter is to build on students' knowledge of two-way tables and Venn diagrams to start to calculate probabilities from these diagrams. This step is a great opportunity to consolidate knowledge of forming and completing two-way tables and Venn diagrams, whilst also extending this knowledge to finding probabilities. When finding probabilities from two-way tables, emphasis should be placed on the key words in the question to help students understand how to decide which total to use. When finding probabilities from Venn diagrams, it is essential that students understand what each part of the Venn diagram represents, in particular the intersection. This chapter also goes on to look at using the product rule for finding the total number of possible outcomes. Here, students are encouraged to reflect on why listing outcomes is not always the most efficient way and to look at how to refine this process to make it more efficient. They are exposed to examples where repeats are allowed and where they are not allowed.

Things to look out for

- Using the wrong total when finding probabilities from two-way tables
- Misinterpreting the question and using incorrect values from any diagram
- Calculation errors when finding missing values in a two-way table or Venn diagram
- Not considering whether repeats are or are not possible when using the product rule

Interleaving opportunities

- Basic probability
- Completing two-way tables
- Completing and interpreting Venn diagrams, including the intersection and union
- Fractions
- Four operations

Models and representations

- Two-way tables and Venn diagrams are used to represent data so that students can find probabilities from them.

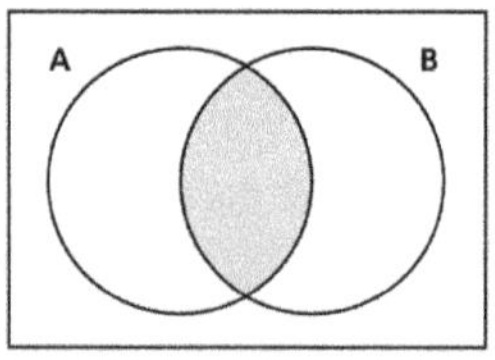

	Owns a pet	Does not own a pet	Total
Year 7			
Year 8			
Total			

Behind the questions

Are you ready?

These three questions provide an opportunity for students to practise the key skills needed in this chapter. Question 1 ensures that they can complete a two-way table and can interpret it by extracting relevant data. Question 2 takes a similar approach with Venn diagrams. In question 3, students consider combinations and outcomes in a context that should be familiar. It is important to encourage them to take a systematic approach and ensure that they have not missed out any of the combinations.

Practice 6.2A

1 Here students should notice that the "Total" column and row are not given on the two-way table. They should copy the two-way table and add these on, as they need to understand the importance of them when finding probabilities. Parts a, b, c and d each require students to use a different total, and students should be able to explain why this is and how to select the correct total to use.

2 When students get to part d, they should recognise that although they can work out the total for Year 11 in the table to be 95, this is not actually the total number of students in Year 11. Instead, it is the total number of students in Year 11 who study either French or Spanish.

3 Here students are required to complete the two-way table before finding probabilities. Watch out for students overlooking the line that says, "They each contain the same number of pencils." This is a crucial piece of information and they will not be able to complete the question without it; however, it is easy to overlook.

4 requires students to find probabilities from a Venn diagram. They should recognise that it is the total number of people in each section that they are interested in. As there is only a small number of pieces of data in this Venn diagram, students could be encouraged to consider each name individually to support their understanding.

5 In part c, watch out for students adding the number in the intersection twice. They should realise that this number does not need to be included twice as it represents the same 14 people who belong in both sets.

6 prompts students to consider "impossible" as a probability. They should recognise that the values given already sum to 50 so there can't be anyone in the intersection.

7 requires students to draw their own two-way table. They should be encouraged to reflect on the benefit of drawing the two-way table and how it assists them in finding the required probability.

What do you think? 6.2A

These two questions get students to think about finding probabilities from diagrams in a slightly different way. Question 1 gets them to think about the two different types of diagrams they have seen and reflect on them both, whilst question 2 looks at some common misconceptions.

Practice 6.2B

1 Here students could be encouraged to draw out the possible combinations and then link this to the abstract calculation to support understanding. As well as knowing that this problem can be worked out by multiplying 5 by 3, it is important that students understand and can verbalise why this is the case.

2 builds on question 1 to ensure that students can explain their reasoning.

3 At this point, students should realise the limitations of listing outcomes and so appreciate the need for the product rule.

4 Watch out for students thinking that Seb's statement is incorrect because they haven't considered the difference between a week and a school week.

What do you think? 6.2B

These two questions get students to think about finding outcomes in a slightly different way. Question 1 gets them to link the number of outcomes to its factors. Questions 2 and 3 look at the idea of replacement and how this affects the number of possible outcomes.

Consolidate – do you need more?

These questions revisit the key points covered in the previous exercises to ensure that students can confidently find probabilities from diagrams.

Stretch – can you deepen your learning?

1 Students should recognise that they need to work out the total value of the coins in order to work out how many 10p coins there are in box B.

2 gets students to reflect on why these probabilities are only estimates and not exact.

3 requires students to think about how slight changes affect the number of possible outcomes.

Reflect

Students should explain their learning in their own words as this will clarify their thinking. They should be able to explain how to find probabilities and outcomes and give examples to support their explanations. They could be stretched to consider where they think people are most likely to go wrong and how to avoid it.

7.1 Forming and using expressions

Small steps

- Form algebraic expressions
- Use directed number with algebra

Rationale

The main purpose of this chapter is to revise the notation and techniques met in Book 1 as groundwork for taking the study of algebra further in this book. The focus is on forming expressions, and a variety of representations are used to support understanding. Simplification is also reviewed, ahead of the upcoming work on expanding one and two sets of brackets, and students are reminded of the notion of equivalence.

The directed number aspect of the chapter aims to develop students' fluency in using expressions of all forms with all numbers. This is a good opportunity to revisit directed number arithmetic, but calculators can be used if this is a barrier to students who are otherwise making progress with their understanding of the algebraic notation.

Things to look out for	Interleaving opportunities
• Students may need reminding of key vocabulary, e.g. "expression", "coefficient", etc. • Errors in simplification, e.g. $2x + 3$ becoming $5x$ – using manipulatives can help students to distinguish between variables and constants • Difficulty with directed number arithmetic – use double-sided counters and number lines for support, and encourage checking using a calculator (students may need reminding how to use the sign-change key)	• Contexts for the use of algebraic notation will occur in almost all of the remaining blocks of the book • Substitution will be a key focus in Blocks 8 and 14

Models and representations

- All the models used in Book 1 are revisited. Algebra tiles, or cubes and counters, are particularly useful for forming expressions and supporting simplification, giving a visual clue as to the difference, for example, between "3 more than x" and "3 times x".
- Function machines are also useful for forming expressions.
- The bar model will be particularly useful to support solving equations and inequalities.
- Some of the expressions that arise could be written using brackets, but this is not essential here as brackets are covered in the next few chapters.

Behind the questions

Are you ready?

The first two questions remind students of basic algebraic notation and substitution. Any issues here will hopefully be covered again through the study of this chapter, but some students may need further intervention. This is also the case for the remaining questions, which focus on directed number. Calculators may be needed to support complex substitution if this remains an issue, so calculator use may need modelling.

Practice 7.1A

1 to **5** provide fluency practice at forming expressions with one or two variables. Provide manipulatives or ask students to draw diagrams for support.

6 uses more formal language, which students may need reminding of.

7 looks at abstract simplification with slightly more complex terms. You could challenge students to find alternative expressions that result in the same answers as for the questions provided. Notice that none of the simplifications in this question needs knowledge of directed number.

What do you think? 7.1A

1 promotes a useful discussion about whether the descriptions are ambiguous and whether they could be reworded and still give the same meaning.

2 links percentages to algebra. Students should be comfortable in writing 25% of x as $\frac{x}{4}$.

3 Encourage students to take a step-by-step approach, looking at the starting variable and how it is manipulated.

4 challenges common errors. Encourage students to use manipulatives and/or substitution to disprove these.

Practice 7.1B

1 and **2** provide fluency practice with substitution of directed number. Number lines, double-sided counters and calculators could all be used as support, as appropriate for your students.

3 is a reminder about the order of operations.

4 uses directed number in the context of simple one-step equations.

5 and **6** look at more complex expressions and simplifications.

What do you think? 7.1B

1 allows students to explore directed number by developing their own expressions.

2 Students may need help to access this investigation. Encourage them to start with $m = 1$ and $n = 1$ and increase the values systematically whilst looking for patterns.

3 uses the same strategy as question 1 but without numerical substitution.

Consolidate – do you need more?

This provides similar questions for further practice, either now or at a later date.

Stretch – can you deepen your learning?

1 can be related to the upcoming study of brackets.

2 looks at simplifying expressions with fractional terms. Students may need a reminder of how to add fractions.

3 can be explored using graphical software to look at powers of x, e.g. $y = -x^4$.

Reflect

This reflection allows students to consider what misconceptions might arise with negative numbers, and how to avoid them.

7.2 Using brackets

Small steps

- Multiply out a single bracket
- Factorise into a single bracket
- Expand multiple single brackets and simplify

Rationale

Multiple representations are used to help students to understand the expansion of brackets, by linking multiplication to repeated addition. In particular, substitution is used to illustrate the equivalence of expressions and also to enable students to recognise that using brackets is a way of recording the mental methods of partitioning for calculations such as 5 × 99.

Factorisation is introduced as "looking for factors". Thinking of factorisation only as "putting into brackets" often leads to initial procedural fluency that is then followed by forgetting the meaning of the word. Nonetheless, it is useful to study expansion and factorisation together so that students make the link between the processes.

Things to look out for	Interleaving opportunities
• Only multiplying the first term within a bracket – using models initially should help to prevent this, as will checking answers by substitution • Incorrect signs – include examples with + and – inside the brackets from the start to prevent this • Incorrect simplification – use the models and methods in Chapter 7.1 for support • Not fully factorising expressions – encourage students to look for the HCF and to consider all the possible ways of factorising an expression such as $12a + 30$	• Simplification and substitution are revisited from the last chapter • Single bracket expansion is reviewed when looking at expanding a pair of binomials, which is the Higher step in Chapter 7.3 • Brackets will feature in the study of equations and inequalities later in this block • Results for sequences involving brackets are used in Chapter 8.2

Models and representations

- Algebra tiles (or cubes and counters as an alternative) can be used to illustrate multiplication as repeated addition. Students can then see the need to find the product of what is outside the brackets and all the terms inside it.
- The area model is a natural development of this without the need to make/draw every single item.
- Both models are equally useful for factorisation as students can consider what possible rectangles can be formed.

Behind the questions

Are you ready?

1 uses function machines to model the difference between $2x + 3$ and $2(x + 3)$, which is key to understanding the content of this chapter.

2 similarly prepares for the multiplication of brackets being seen as repeated addition.

3 and **4** involve the key skills of expanding and factorising, respectively.

Practice 7.2A

1 This fluency question can be supported by the use of algebra tiles.

2 and **3** make the links between the abstract use of brackets and mental calculation methods.

4 and **5** provide further fluency practice. The area model could be used for support here.

6 looks at repeated application of the key skill. Simplification is also required.

What do you think? 7.2A

1 tackles the common difficulty of "– (x + a)". Encourage students to think of this as "– 1(x + a)".

2 reinforces the links to mental methods.

Practice 7.2B

1 Dividing terms can be seen as more difficult than multiplying them, so this is looked at first.

2 Use the word "factors" when supporting students here and as a lead in to question 3.

4 Using manipulatives may help students to see the highest common factors.

5 and **6** look at more complex expressions. Again, encourage students to look for the HCF.

What do you think? 7.2B

1 Students may need support to see the link with factorising, especially in part d, where you may need to remind them that 24 is a factor of 48.

2 is more abstract, but can be linked back to question 1.

3 emphasises the need for a common factor.

4 Encourage students to compare coefficients.

Consolidate – do you need more?

This provides similar questions for further practice, either now or at a later date.

Stretch – can you deepen your learning?

1 provides a context where repeated expressions may occur and challenges students to find their own.

2 looks at more complex working.

3 extends the earlier idea of factorising numerical calculations to include decimals.

Reflect

The reflection encourages students to make and remember the link between expanding and factorising, and finding the highest common factor of terms.

7.3 More than one set of brackets

Small steps

- Expand a pair of binomials (H)
- Expand multiple single brackets and simplify

Rationale

The focus of this short chapter is the Higher step of expanding two binomials. If you think your students are not ready for this, then you can omit it as it will be covered again in Book 3. The common approach of "FOIL" is not included as the chapter concentrates on developing conceptual understanding rather than just learning a rule. Building on Chapter 7.2, expansions are found using multiple representations, including algebra tiles and the area model, before progressing to the purely abstract form. You may wish to concentrate on the less formal representations this year and move on to the more abstract form next year if this is appropriate for your class.

Making links with arithmetic, both in terms of demonstrating the expansion and being able to check answers through substitution, is encouraged. Factorisation of quadratics is not included.

Things to look out for	Interleaving opportunities
• Errors in finding the terms – stay with the area model, showing all four terms clearly if students are making errors with other forms, and show the different models alongside each other to help students to appreciate the similarities • Difficulty with complex coefficients – stay with easier numbers until students are confident before moving on to explore expansions of the form $(ax + \ldots)(bx + \ldots)$ • Errors with directed number – use algebra tiles and zero pairs to help • Incorrect simplification – again, use real models to emphasise like and unlike terms	• This chapter builds on the learning from Chapters 7.1 and 7.2 • You could link the expansion of two binomials to generating sequences from complex rules in Chapter 8.2 • There are few other natural links in the remainder of this book, so you could keep the skill sharp by revisiting it in starters

Models and representations

- The area model is the most appropriate representation when the binomials feature two variables, and this can be used alongside algebra tiles when looking at binomials with the same variable.
- Arrows are used to show the multiplication of terms in the more abstract representation, linking to the model used in Chapter 7.2.

Behind the questions

Are you ready?

1 checks that students are comfortable with expanding a single bracket.

3 builds on this to check that students can form terms involving more complex multiplications.

2 and **4** look at the simplification that will be needed for binomials with the same variable.

Practice 7.3A

1 to **3** build on the use of the area model from numbers to variables. Drawing and substitution are to be encouraged.

4 and **5** move on to expansions with a single variable where simplification is needed. Algebra tiles and the area model can be used for support here, and students will need to be confident with directed number. Substitution is again a useful check, and can be linked to individual parts of area model diagrams.

What do you think? 7.3A

1 As an extension, this looks at expansions where the coefficient of the variable is greater than 1 in at least one bracket. Tiles and area models can still be used to support these more complex expansions.

2 tackles a common misconception around squaring and then provides fluency practice with this.

Consolidate – do you need more?

This starts with the link to long multiplication before providing further fluency practice.

Stretch – can you deepen your learning?

1 is an investigation into the difference of two squares. This can also be illustrated diagrammatically.

2 looks at comparing coefficients to find missing values.

3 encourages students to make connections between the numbers in the binomials and the result. You could challenge them by linking this to some simple factorisation, but this is not needed at this stage.

4 challenges students to think about how to expand three binomials.

Reflect

The reflection encourages students to consider how the method for expanding two binomials works. They could illustrate their answers with diagrams.

7.4 Brackets and equations

Small steps

- Solve equations, including with brackets
- Form and solve equations with brackets

Rationale

As students have extensive experience of solving one-step and two-step equations from Book 1 and within the contexts of earlier chapters in this book, they will now move on to exploring equations with brackets. This provides an opportunity to practise their skills from Chapter 7.2 as well as to revise solving equations. Many equations with brackets can be solved by dividing first rather than expanding, for example by starting with division when there is a common factor, so both methods are explored and students can choose which they prefer.

As in Book 1, equations are presented in a variety of forms and not just as $a(bx + c) = d$. The chapter also explicitly looks at forming equations, using number puzzles and other areas of mathematics as contexts.

Things to look out for

- Errors in expanding brackets as met in Chapter 7.2 – alert students to potential pitfalls and use this chapter as an opportunity to learn from any earlier mistakes
- Expecting (positive) integer solutions – include all sorts of answers in your modelled examples
- Trying to "spot" answers – as above, choose examples where the answers are not obvious and encourage students to use their mathematical knowledge to work towards solutions
- Transcription errors are common in increasingly longer solutions – encourage students to work in pairs to check each other's working to identify and eliminate errors

Interleaving opportunities

- The chapter builds on the learning from Chapters 7.1 and 7.2, and will influence the learning of the rest of the block
- Area, perimeter and angles are included as contexts for forming and solving equations, but you could add in examples from almost any area of mathematics that you feel students would benefit from revising
- Equations will feature in many of the forthcoming chapters, especially in Blocks 8, 14, 15 and 17

Models and representations

- All the familiar models from Book 1 are included to help students to "see" equations and how they might be solved. The bar model, in particular, provides a clear visual sense of both multiplying out brackets and collecting like terms to maintain equality, and can be used to demonstrate solving equations by dividing first.
- Function machines or flowcharts can also be used for "working backwards".

Behind the questions

Are you ready?

The first three questions assess how well students have remembered how to solve one-step and two-step equations. If there are problems here, then more revision of these will be needed before proceeding to the main part of the chapter; there are plenty of practice questions in Book 1.

Likewise, question 4 checks the necessary fluency with expanding brackets, which has just been covered in the previous chapters.

Practice 7.4A

1 compares solving an equation with and without brackets as a discussion point.

2 The parts can all be approached by first dividing by the number outside the brackets or, if preferred, by first expanding the brackets. Compare these methods as a class.

3 and **4** look at more complex coefficients. Look out for accuracy errors here.

5 involves brackets and fractional terms.

6 introduces negative coefficients of x. This will be dealt with again when working with inequalities.

What do you think? 7.4A

Both questions look at number puzzles as a vehicle for introducing the ideas of the next section. You could support students by using representations such as function machines or flowcharts.

Practice 7.4B

1 and **2** involve simple expressions and equations.

3 and **4** are useful for promoting discussion about the choice of variable when there are linked unknowns.

5 and **6** You may need to remind students about perimeter and the names of shapes.

What do you think? 7.4B

1 is slightly more complex. Refer back to question 3 in Practice 7.4B, where the choice of variable was considered.

2 interleaves angle facts. Students may need a reminder of notation and properties.

Consolidate – do you need more?

This provides similar questions for further fluency practice, either now or at a later date.

Stretch – can you deepen your learning?

1 introduces more steps. Compare multiplying both sides of each equation by the denominator first, to clear the fraction, with expanding the brackets by multiplying by the fractions.

2 and **3** are an exploration of the material coming in Chapter 7.6. Encourage a visual approach rather than trial and improvement.

Reflect

These reflections could form the basis of a display looking at alternative approaches to forming, solving and checking solutions to equations.

7.5 Inequalities

Small steps

- Understand and solve simple inequalities
- Form and solve inequalities

Rationale

A lot of time is invested in the early part of this chapter to ensure that students understand the meaning of the symbols used when working with inequalities, particularly as $\geqslant$ and $\leqslant$ may be new to them. The balance model, which should be familiar from previous work on solving equations, is used to model inequality and to illustrate that, in many ways, an inequality is similar to an equation. It is useful to build up relationships such as $y > 4$ implying that $y + 1 > 5$ before reversing the process to solve inequalities.

Care is needed to ensure that students don't think that inequalities can always be treated in exactly the same way as equations because multiplying/dividing inequalities by a negative number reverses the direction. This is explored in detail in Book 3. Solutions in the form of sets (e.g. $x > 17$) and the greatest/least integer solutions are both looked at again to consolidate students' understanding of the situations and the solutions.

Things to look out for	Interleaving opportunities
• Treating an inequality exactly as an equation and changing/omitting the sign • Model solving inequalities carefully and discuss the meaning of solution sets, in particular to avoid correct working followed by $x = \ldots$ rather than $x > \ldots$, say • Not appreciating that $x > 3$ and $3 < x$, say, are equivalent – look at alternative forms and test them by substituting several values • Difficulty with the $\geqslant$ and $\leqslant$ symbols – discuss situations where the corresponding phrases "at least …" and "at most …" are appropriate	• As well as algebraic notation from Chapter 7.1, there are clear links with solving equations from Chapter 7.4 • As with the previous chapter, you could add examples from almost any area of mathematics that you feel students would benefit from revising • Forming and solving inequalities is explicitly featured here in Block 7, but the notation could be used, discussed or revised in almost all contexts in the rest of the book • Non-integer solutions could be presented as decimals, or you could revise conversion to mixed numbers

Models and representations

- The balance is a great model for illustrating quantities that are not equal, but students may have difficulty reading and recording which value is greater, so lots of modelling and discussion are useful.
- A bar model can be used in a similar way to when solving equations, with one bar being longer/shorter as appropriate, but $\geqslant$ and $\leqslant$ would need some discussion.

Behind the questions

Are you ready?

1 looks at ordering as a skill check before question 2 tests recall of the > and < symbols. You may need to provide intervention to help students work with directed number. Simple equations that correspond to forthcoming simple inequalities are then reviewed.

Practice 7.5A

1 to **4** focus on understanding different representations. Students often find inequality confusing, and it is worth investing time in class discussion to clarify any misconceptions that may arise.

5 provides simple fluency practice.

6 extends to looking at greatest/least integer solutions.

What do you think? 7.5A

1 develops finding integer solutions whilst reminding students of the value of π.

2 tackles the problem of multiplying an inequality by a negative number, and illustrates the correct approach as an alternative. This is explored in detail in Book 3.

Practice 7.5B

1 extends to the use of brackets.

2 to **7** provide simple contexts in which students can form and solve inequalities. Question 5 is revisited in Block 8.

What do you think? 7.5B

1 explores the misconception that all solutions are integers.

2 introduces the notation for a number within a range. You could use number lines to illustrate this if you wish.

3 links inequalities to the side lengths of triangles.

Consolidate – do you need more?

This provides similar questions for further fluency practice, either now or at a later date, again starting with an understanding of the notation.

Stretch – can you deepen your learning?

1 Students may need to be guided towards considering negative numbers.

2 Students may need prompting to realise that both inequalities need to feature the "+ 3".

3 increases the level of difficulty of the arithmetic. Encourage students to multiply to avoid working with fractions; you could compare both ways if you prefer.

Reflect

Students' reflections should include the need to be consistent with symbols. Also, it might be worth noting that $3 = x$ and $x = 3$ are equivalent but $3 > x$ and $x > 3$ are not.

7.6 Unknowns on both sides (H)

Small steps

- Solve equations and inequalities with unknowns on both sides (H)
- Form and solve equations and inequalities with unknowns on both sides (H)

Rationale

This is a natural extension of the material in the other chapters of this block, but should only be tackled when students are ready. The work will be revisited in Book 3 if students need more time to consolidate their ability to solve easier equations this year. Alternatively, you could study the first part of the chapter but omit the more challenging second part.

Equations with three and four terms are included as students who are procedurally fluent at solving equations of the form $ax + b = cx + d$ can be stumped when presented with those of the form $ax + b = cx$. Bar models and/or balances are particularly useful for developing understanding of both three- and four-term equations.

The chapter explores solving equations and inequalities before moving on to forming them, but you could look at vehicles such as number puzzles to produce equations first if you wish.

Things to look out for	Interleaving opportunities
• "Losing" negative signs – encourage students to work with letters before numbers, as working the other way round is when this problem is most likely to occur	• This chapter develops the material in the last few chapters • Angle rules and perimeter are explicitly explored, but you could add examples from almost any area of mathematics that you feel students would benefit from revising

Models and representations

- The balance is again a useful model for looking at unknowns on both sides, as additions/subtractions of both the algebraic and constant terms can be easily illustrated. It is not as obvious for representing equations with negative coefficients, although different colours (akin to algebra tiles) could be used. Using this model alongside an abstract method of solving simple equations should help students to become familiar with the method before moving to more complex equations in the abstract only.
- The bar model is also a suitable model, and it has the benefit of being easy for students to draw in their books.

Behind the questions

Are you ready?

These questions assess whether students are proficient at solving a variety of equations with unknowns on one side, both with and without brackets, before proceeding to this more challenging Higher step. If there are issues, you may wish to leave step this until students study Book 3.

Practice 7.6A

1 and **2** look at models and representations to support students when solving this new type of equation.

3 explores the different types of three- and four-term equations and can be used as a starting point for a useful discussion.

4 and **5** provide fluency practice.

6 extends to equations involving brackets.

What do you think? 7.6A

1 and **2** introduce forming equations with unknowns on both sides in the context of number puzzles.

3 and **4** illustrate that inequalities with unknowns on both sides can be approached in the same way as equations.

Practice 7.6B

1 extends to equations and inequalities with terms that have negative coefficients. Students need to be confident with the abstract approach to solve these.

2 to **5** provide simple contexts in which students can form and solve equations with unknowns on both sides.

What do you think? 7.6B

1 revisits three-term equations.

2 and **3** explore the use of fractional terms.

Consolidate – do you need more?

This provides similar questions for further fluency practice, either now or at a later date. The focus is on equations, but questions could be altered to discuss inequalities if you wish.

Stretch – can you deepen your learning?

These questions introduce the idea of simultaneous equations, using a visual approach. This is just for exploration and there is no intention of studying these equations in depth at this stage.

Reflect

Students reflect on their approaches. Encourage them to include examples of all types of equations, with positive and negative coefficients.

7.7 Making sense of algebra

Small steps
• Identify and use formulae, expressions, identities and equations

Rationale

Having explored all sorts of algebraic constructs throughout this block, this final chapter cements students' understanding of the vocabulary associated with these. A particular focus is that a formula can be used to generate equations which can then be solved. This will be considered again in Book 3 when students will explore changing the subject of a formula.

Students should be encouraged to use the correct vocabulary when they come across algebra in other areas of mathematics throughout the remainder of the book.

Things to look out for	**Interleaving opportunities**
• Thinking that any mathematical statement with an equals sign is an equation – discuss how formulae differ from equations and how they can be used to generate equations • Giving incomplete formulae – discuss the difference between an expression for a perimeter such as $2l + 2w$ and a formula such as $P = 2l + 2w$ • Mixing up equations and identities – reinforce that an identity is true for *all* values of the variables, and illustrate this by substituting a wide range of values	• Use of brackets is revised • This language could be revisited whenever these constructs occur throughout the book

Models and representations

- There are no new models or representations in this chapter because it focuses on understanding the algebraic symbols used to create the constructs.

Behind the questions

Are you ready?

These questions use the vocabulary of the chapter and should highlight any gaps in students' understanding of the definitions.

Practice 7.7A

1 provides comparison between formulae and equations.

2 revises the use of the identity symbol.

3 and **4** look at using formulae to produce equations.

5 and **6** introduce creating formulae before using them to form and solve equations.

7 looks at comparing coefficients in an identity.

What do you think? 7.7A

1 is a useful discussion point as the phrase "equation of a straight line" can be confusing!

2 looks at equivalent forms of the same expression. You could also use the expressions to create formulae.

3 explores a more complex formula.

Consolidate – do you need more?

This provides similar questions for further fluency practice, either now or at a later date.

Stretch – can you deepen your learning?

1 and **2** look informally at the idea of changing the subject of a formula, which is studied in Book 3.

3 and **4** look at slightly more complex formulae.

Reflect

Students are challenged to show their understanding of the terminology. This could form the basis of another useful display.

8.1 Revisiting sequences

Small steps
• Generate sequences given a rule in words • Generate sequences given a simple algebraic rule

Rationale

Students studied sequences in Book 1, and this chapter revises that material, focusing on sequences given in words and on simple algebraic rules. The language of sequences, e.g. linear, geometric, Fibonacci and term-to-term rule, is revisited. Although more abstract than the material in Book 1, students could still use concrete and pictorial representations to see the behaviour of sequences.

Algebraic sequences in this chapter are limited to the form $an + b$ to help students to become familiar with this way of representing a linear sequence; more complex rules are explored in the next chapter.

Things to look out for	**Interleaving opportunities**
• Making incorrect generalisations, such as thinking that the tenth term is double the fifth term – demonstrate that this can be true, but is not always true (this is explored within the chapter). You could link this to the structure of the sequence or the algebraic representation, if appropriate. • Assuming that all sequences are linear – model both linear and non-linear sequences	• The rules for sequences are used to form equations and inequalities, revisiting the last block • The learning is taken further in the next chapter • You can practise non-calculator skills and the rules of directed number

Models and representations

- Sequences can be represented in many ways, including lists, tables and graphs.
- Using cubes, counters or diagrams to illustrate the terms of a sequence can help students to spot the pattern and also the underlying structure.

Behind the questions

Are you ready?

The questions assess how well students have remembered the language of sequences from Book 1.

Practice 8.1A

1 to **3** look at generating and continuing sequences given in a variety of forms, to reinforce the language and, in particular, the notion of a term-to-term rule.

4 moves on to describing sequences fully, and emphasises the need for giving the starting number as well as the term-to-term rule.

6 to **8** explore more complex rules for getting from one term to the next, including those presented in the form of a flowchart. Students may need support to see the common factor of 3 in question 8.

What do you think? 8.1A

1 and **2** look at variations of the Fibonacci sequence. You might need to help students to spot the pattern. This could be extended to start with negative numbers so as to practise directed number arithmetic.

3 looks again at different types of sequences. Students may need support when phrasing rules where the term-to-term rule changes. Working in pairs on this might be useful.

Practice 8.1B

1 revises substitution into a simple rule of the form $an + b$. The link between the coefficient a and the constant difference is touched on in preparation for Chapter 8.2.

2 and **3** look at using larger values of n and include a decreasing linear sequence. Students could check their arithmetic on a calculator.

4 and **5** interleave forming and solving equations and inequalities. Students may need a reminder of these topics.

6 explores last digits. You could model this by looking at $10n + a$ first as a scaffold.

What do you think? 8.1B

1 can be supported using cubes and counters to explore $2n$ and $2n + 1$.

2 helps students to understand that, for example, the tenth term is double the fifth term when the rule is of the form an, but not when the rule is $an + b$ for non-zero b. Linking this to times tables and concrete representations would be useful.

Consolidate – do you need more?

The focus of the consolidation questions is substitution and the vocabulary of sequences, but forming and solving equations is included too. Students could also make up their own similar questions and challenge each other for further practice.

Stretch – can you deepen your learning?

1 looks at the idea of generalising too early. Extend this problem, asking students to create their own sequence that could be of a different form.

2 and **3** might be approached by trial and improvement, but you could model this by setting up and solving equations.

4 is just for fun!

Reflect

The focus is on generalisation and being comfortable with the large amount of vocabulary.

8.2 Moving on with sequences

Small steps

- Generate sequences given a complex algebraic rule
- Find the rule for the nth term of a linear sequence (H)

Rationale

Students should now be confident enough with substitution to explore sequences given by more complex rules including squared and cubed terms or using brackets. They may wish to expand the brackets, so you could have a discussion about why this isn't necessary, by substituting into both forms and comparing the answers.

If appropriate, students can go on to study the Higher step of finding the rule for the nth term of a linear sequence, although this is revisited in Book 3 if students are not yet ready. The approach taken is to compare sequences with the relevant times table and adjust rather than using a procedural approach and the confusing abstract notion of a "0th term".

Things to look out for

- The absence of an explicit n^2 term in rules like $n(n + 1)$ may lead students to think that this is linear – demonstrate by substitution and expansion that this is not the case
- Likewise, believing that $\frac{1}{n}$ is linear as there is no squared term – again, this is easily disproved by substitution
- Mistakes in the signs when forming rules for the nth term are common, e.g. writing $4n – 3$ instead of $4n + 3$ – encourage students to check their answers by substituting at least two terms (this can be done mentally)
- Thinking that the rule for a sequence such as 6, 7, 8… is $n + 1$ – checking the first term dispels this

Interleaving opportunities

- This chapter builds on Chapter 8.1
- Bracket notation from Block 7 is revisited
- Arithmetic skills and square numbers
- Substitution into formulae is developed, including fractional algebraic notation
- Order of operations

Models and representations

- Students are encouraged to make or draw the terms in non-linear sequences to see that there is no constant difference. Making and drawing sequences with an n^2 term (or equivalent) shows the rapid growth and the need for abstract representation.
- Students should make the link between the use of algebra and concrete representations, for example linking the rule $2n + 1$ for this pattern sequence to its pictorial representation by thinking of it as "2 more are added every time" and then "1 more is needed to make the first triangle".

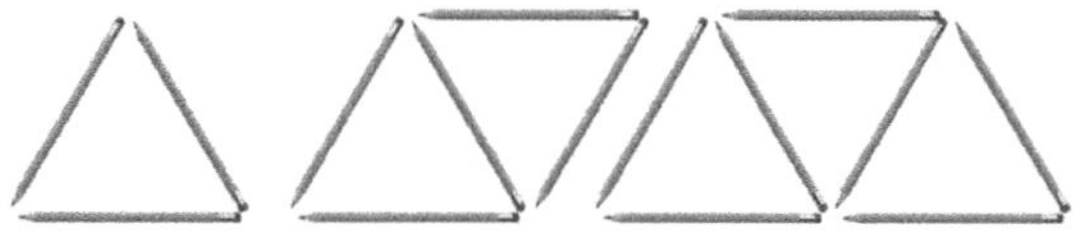

Behind the questions

Are you ready?

The first two questions prepare students for the substitution needed in this chapter, and any difficulties will mean that intervention is necessary before proceeding. Question 3 looks at the order of operations for more complex substitution. Question 4 is a preparation for the second step, so could be omitted if you don't intend looking at this with your class.

Practice 8.2A

1, **2** and **4** provide fluency practice with substitution into more complex expressions, with question 4 being the most prone to error and therefore a good one to discuss as a class.

3 requires students to verify a result, and this may need some guidance.

5 checks whether students can recognise linear rules of the form $an + b$. To highlight the misconception that any sequence whose rule does not involve n^2 is linear, sequences of the form $\frac{1}{n}$ and $n(n + 5)$ are included. Students could think of other similar rules that don't give linear sequences and challenge each other to recognise each type.

6 and **7** look at yet more different types of rules to develop students' substitution skills. It is worth discussing the similarities between the sequences.

What do you think? 8.2A

1 and **2** deepen the exploration by looking at sequences of fractions where the numerators and denominators each follow a separate rule. Students should avoid simplifying to the lowest terms as this can make the patterns less obvious. Some students may need to convert the fractions to decimals to appreciate the long-term behaviour of these sequences.

Practice 8.2B

1 and **2** make explicit the connection between a in $an + b$ and the a times table and the fact that b does not affect this.

3 links these together and guides students to finding the nth term rule when a is already given.

4 removes some of the scaffolding so that students are ready to proceed independently by the time they reach question 5. However, you could provide scaffolding here if necessary.

6 explores a common misconception.

7 links the use of algebra with concrete representations of sequences. If you prefer, you could start with this type of exploration instead.

What do you think? 8.2B

1 looks at negative coefficients of n. Students need to be confident with directed number arithmetic before they attempt this section. Finding the rule for decreasing linear sequences can be challenging, so encourage students to use substitution to check that they have made the correct choice.

2 looks at rules for linear sequences given in unexpected forms.

3 explores second differences in quadratic sequences. This is an extension and is designed to prevent the misconception that the method for finding a rule in the form $an + b$ cannot be used for all sequences. Finding the rule for the nth term of a quadratic sequence is not expected at KS3.

Consolidate – do you need more?

This is mainly fluency practice for use as homework. Alternatively, it can be revisited at a later date.

Stretch – can you deepen your learning?

1 encourages students to recognise sequences similar to n^2.

2 looks at finding rules and builds on the first "What do you think?" section. Encourage students to look at the numerators and denominators separately.

Reflect

The reflections here cover the main points of the learning and could be recorded either in journals or in the form of a display. As this is a relatively stand-alone topic, it may be worth encouraging students to return to these reflections at a later date to support retention.

9.1 Understanding indices

Small steps

- Add and subtract expressions with indices

Rationale

The purpose of this chapter is to recap prior learning on collecting like terms but with a specific focus on expressions that contain indices. Students often mix up adding and subtracting terms with indices, and mistake any term containing x, e.g. x^4 and x^2, as being like terms. The purpose of this chapter is to ensure that students can not only add and subtract expressions with indices, but they also understand when they can and when they can't simplify an expression. The use of algebra tiles can support basic understanding of why x and x^2 are not like terms. For example, if you have four x tiles and eight x^2 tiles, it doesn't mean that you have twelve x^3 tiles. Using algebra tiles to illustrate $2x^2 + 2x^2$ can support understanding of why the coefficients are added but the powers are not.

Things to look out for	Interleaving opportunities
• Writing 1 as a coefficient • Adding/subtracting the indices when simplifying expressions • Combining terms that have the same index but different bases, e.g. thinking that $x^2 + y^2 \equiv xy^2$	• Area • Perimeter • Algebraic notation • Like terms • Fractions • Decimals

Models and representations

- Algebra tiles are a good way of representing expressions that involve indices.

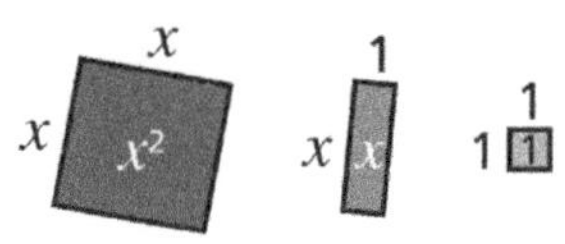

- For example, here are one x^2 tile, three x tiles and two "1" tiles, so the expression is $x^2 + 3x + 2$ not $6x^3$.

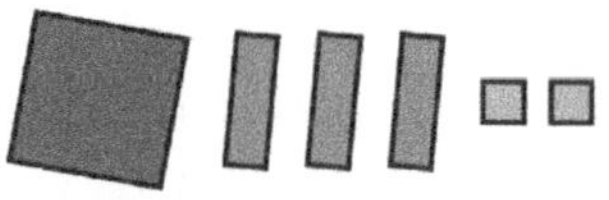

Behind the questions

Are you ready?

These three questions cover the essential knowledge for this chapter. Question 1 looks at basic calculations with directed number to ensure that these don't cause a barrier to progress within the exercises. Question 2 ensures that students understand what is meant by "like terms", and question 3 builds on this by asking students to collect like terms in basic expressions.

Practice 9.1A

1 uses algebra tiles to support students' understanding of simplifying expressions with indices. They should recognise that the numbers of tiles in each part are the same, but the types of tiles used are different. This can be used to highlight why 6, $6x$ and $6x^2$, for example, are not like terms.

2 addresses a common misconception. Students could be encouraged to use or draw algebra tiles to support their reasoning.

3 Students start to simplify expressions with indices but they are still working with the aid of a diagram. Describing each diagram in words can support their final answer: for example, "Two x^2 tiles and three x tiles can't be simplified any further", so the answer to part a is simply $2x^2 + 3$.

4 prepares students for moving into the abstract in question 5.

5 Here students work abstractly to simplify expressions that contain only squared terms. Students should be able to explain why all the terms in each expression are like terms. In parts a to d, x continues to be the variable used, but this changes in later parts of the question. Algebra tiles could be used to support students if necessary. They should be able to quite quickly generalise that because all the terms are like terms, you add/subtract the coefficients and the power stays the same.

6 and **7** are similar to questions 4 and 5 but this time with a focus on cubic terms.

8 looks again at the common mistake of collecting terms that are unlike. Students should be able to use the diagram to explain why the two types of objects are not the same, and then explain why the expression cannot be simplified further.

9 requires students to identify like terms before simplifying each expression. All the powers are 1, 2 or 3 so cubes and algebra tiles can be used for support if they are needed.

10 and **11** build on question 9 by moving on to higher powers. Students can no longer rely on manipulatives. Again, they should first consider which terms are like and why before collecting them to simplify the expressions.

What do you think? 9.1A

These three questions are designed to make students think more deeply about expressions with indices. In question 1 they focus on which terms are like and which aren't. Watch out for students thinking that the inclusion of fractions or negatives complicates the question. They should recognise that the rationale behind what makes terms like or unlike is unchanged. Question 2 addresses the misconception that all terms with the same power are like terms. Question 3 introduces an element of problem solving by asking students to write an expression that simplifies to a given one.

Consolidate – do you need more?

These questions revisit the key points covered in the previous exercise, and involve simplifying expressions with indices. The use of manipulatives can support students if needed. If more practice is required, students could work in pairs to produce similar questions and challenge each other.

Stretch – can you deepen your learning?

1 In part b, students might think that this isn't possible as they're not like terms. However, their attention should be drawn to the word "expression". The question is specifically asking about expressions, and not terms, and so it is possible to find an answer.

2 Here the expressions to be simplified involve fractions. Students should recognise that in part a the expressions on each side of the equals sign represent the same thing – they are just written differently – and then use this knowledge to help them in part b.

3 looks at more complex expressions with indices. Watch out for students thinking that x^2y and xy^2 are like terms.

Reflect

Students should explain their learning in their own words as this will clarify their thinking. They should be able to identify like terms and explain clearly what makes two terms like.

9.2 Multiplying expressions

Small steps

- Simplify algebraic expressions by multiplying indices
- Use the addition law for indices

Rationale

Here students are introduced for the first time to the addition law for indices. They start by simplifying algebraic expressions involving repeated multiplication by using indices, which will help them to understand what is meant by a power. They then go on to explore what happens when two terms with powers are multiplied, by writing them in expanded form. Students often want to move to a very procedural method quite quickly but they should be encouraged to show working and explain their reasoning throughout to ensure that they understand what they are doing rather than just knowing how to work out the answer.

Things to look out for	Interleaving opportunities
• Adding the indices when two terms don't have the same base • Multiplying the indices rather than adding • Adding the coefficients rather than multiplying • When the base is a number, multiplying the numbers when the base should remain unchanged	• Four operations • Algebraic notation • Solving equations • Expanding brackets

Models and representations

- Being able to represent numbers and letters with powers in expanded form is essential for understanding the addition law for indices and is useful when unpicking misconceptions. For example, 2^4 in expanded form is $2 \times 2 \times 2 \times 2$ and y^3 in expanded form is $y \times y \times y$.

Behind the questions

Are you ready?

These two questions cover the essential knowledge. It is important that students understand what is meant by a power as they will build on this throughout the chapter. In question 2, students are reminded of prior learning on algebraic notation and that the multiplication symbol isn't written between two letters or between a letter and a number.

Practice 9.2A

1 Here students are given the working and the answer and have to carefully explain the steps that have taken place. In part b, they apply this process for themselves. Again the answer is given, so the focus is not on the answer but on how to get there and the thinking along the way. Students should notice that the powers are added, but they don't need to generalise this yet.

2 gives students the opportunity to practise simplifying expressions. Initially, they should be encouraged to write out the answers in full using expanded form, but they might want to move on from this quite quickly. Even then, they should be able to explain clearly how they've arrived at their answers. In part b, if students are working abstractly, watch out for them not recognising that y is y to the power 1.

3 allows students to practise simplifying expressions with indices. They should take care with parts d and e, where the base is a number. Students should realise by now that the base stays the same and the powers are added, and they should apply this logic here. Some may be tempted to multiply the numbers, but they should be prompted to recognise why this is incorrect by writing the expressions out in full.

4 introduces coefficients. Because they have been adding the powers, students may think that they also need to add the coefficients. Again, students could be encouraged to write the calculation in expanded form to support their understanding of why this is incorrect.

5 allows students to consolidate using the addition law for indices to simplify expressions with coefficients.

6 requires students to find missing powers. They should be encouraged to write down the steps in their thinking rather than just find the answer.

7 This may appear more challenging than the other questions. However, students should be prompted to explain the steps they need to take and then use prior knowledge of simplifying expressions to help them.

8 ensures that students know that the base needs to be the same in order to apply the addition law for indices.

What do you think? 9.2A

These three questions are designed to make students think deeply about the addition law for indices.

1 exposes common misconceptions from this topic.

2 asks students to identify terms that give a required product. They could be stretched to generalise what must be true in order for the two terms to have the given product.

3 looks at this topic in relation to number. Watch out for students just trying to find the answer as it isn't algebraic; instead, they should focus on what the base must be and then ensure that all the numbers in the product are written with that base before simplifying.

Consolidate – do you need more?

These questions revisit the key points covered in the previous exercise and give students further practice at using the addition law for indices. Students should be encouraged to write each calculation in expanded form to support their thinking. If more practice is needed, students could be asked to work in pairs to produce similar questions and challenge each other.

Stretch – can you deepen your learning?

1 builds on the final question from the "What do you think?" section by looking at powers of numbers. Watch out for students thinking that 27 is 9 to the power 3 because 9 multiplied by 3 is equal to 27.

2 provides some more challenging examples of simplifying expressions. Students may be tempted to expand the brackets in parts k and l but, instead, they should be prompted to notice that as the base is the same, they can simply add the indices.

3 Some students may try to use trial and improvement to work out the missing values. Instead, they should form and solve equations and be able to explain each step in their working.

4 interleaves prior learning and requires students to expand brackets before simplifying the expressions. Watch out for students not simplifying correctly.

Reflect

Students should explain their learning in their own words as this will clarify their thinking.

9.3 Dividing expressions

Small steps

- Simplify algebraic expressions by dividing indices
- Use the addition and subtraction law for indices

Rationale

Here students are introduced for the first time to the subtraction law for indices. They start by simplifying some algebraic expressions involving indices. They then go on to explore what happens when two terms with powers are divided, by writing them in expanded form, which will help them to understand the subtraction law. Students often want to move to a procedural method quite quickly, but they should be encouraged to show working and explain their reasoning throughout to ensure that they understand what they are doing rather than just knowing how to work out the answer. Students will continue to use the addition law for indices. Again, they should not be too procedural and should be encouraged to explain their thinking.

Things to look out for	Interleaving opportunities
• Adding/subtracting the indices when two terms don't have the same base • Multiplying/dividing the indices rather than adding/subtracting • Adding/subtracting the coefficients rather than multiplying/dividing • When the base is a number, multiplying or dividing the numbers when the base should remain unchanged	• Four operations • Algebraic notation • Solving equations • Expanding brackets

Models and representations

- Being able to represent numbers and letters with powers in expanded form is essential for understanding the addition and subtraction laws for indices and is useful in unpicking misconceptions. For example, 2^4 in expanded form is $2 \times 2 \times 2 \times 2$ and y^3 in expanded form is $y \times y \times y$.

Behind the questions

Are you ready?

These three questions cover the essential knowledge for this chapter. Question 1 recaps what was covered in Chapter 9.2. Question 2 focuses on division, presented both using the symbol and as a fraction. Question 3 looks at simplifying fractions, which will be used later in the chapter when simplifying expressions.

Practice 9.3A

1 In part a, students are given the working and the answer and they need to carefully explain the steps that have taken place. In part b, they then apply this process for themselves. Again, the answer is given so the focus is not on the answer but instead on how to get there and the thinking along the way. Students should notice that the powers are subtracted, but they do not need to generalise this yet.

2 exposes a common misconception and gives students the opportunity to unpick it.

4 Here students practise simplifying expressions. Initially, they should be encouraged to write out their answers in full using expanded form, but they might want to move on from this quite quickly. Even then, they should be able to explain clearly how they arrived at their answers. In part b, if students are working abstractly, watch out for them not recognising that *y* is *y* to the power 1.

5 gives students the opportunity to practise simplifying expressions with indices. Watch out for parts c, e and k, where the base is a number. Students should realise by now that the base stays the same and the powers are subtracted and should apply this logic here. Some may be tempted to divide the numbers, but they should be prompted to recognise why this is incorrect by writing the expressions out in full.

6 introduces coefficients. Because students have been subtracting the powers they may think that they also need to subtract the coefficients. Again, students could be encouraged to write the calculation out in expanded form to support their understanding of why this is incorrect.

7 also includes coefficients but this time they are written as fractions. Students often find this hard. They could be encouraged to write these in full using division or to think about the process as simplifying fractions. Students should be encouraged to notice the similarities and differences between each pair.

9 combines the addition and subtraction laws for indices. Watch out for students dividing both terms on the numerator by the denominator. If this happens, they should be encouraged to write out the expressions in full to realise why this is incorrect.

What do you think? 9.3A

These five questions require students to think deeply about the subtraction law for indices. Question 2 asks students to identify terms that divide to give a required answer. They could be stretched to generalise what must be true in order for the two terms to give the required result when divided. Question 3 exposes the misconception that the base changes when it is a number. Question 4 introduces the idea of the power 0 without explicitly formalising it. Students should already be aware that if the numerator of a fraction is equal to the denominator, then the fraction is equivalent to one whole. In question 5, watch out for students trying to calculate the answer rather than using the given information to rewrite and simplify the calculation.

Consolidate – do you need more?

These questions revisit the key points covered in the previous exercise, and give further practice at using the addition and subtraction laws for indices. Students should be encouraged to write each calculation in expanded form to support their thinking. If more practice is needed, students could be asked to work in pairs to produce similar questions and challenge each other.

Stretch – can you deepen your learning?

1 looks at the relationship between the powers. Students should recognise that in part a x must be 3 more than y and that in part b x must be 1 more than y. Before answering part b ii, students could look back at question 7 in Practice 9.3A to help them to understand why if $x - y$ is 1, then $y - x$ is -1.

2 and **3** provide some more challenging examples of simplifying expressions.

4 provides practice at the subtraction law for indices, with variables in the exponents.

5 Students generalise their learning from the previous two chapters and explicitly form the addition and subtraction laws for indices. They could be encouraged to substitute values for a and b to test their answers.

Reflect

Students should explain their learning in their own words as this will clarify their thinking.

9.4 Powers of powers (H)

Small steps

- Explore powers of powers (H)

Rationale

Here students are introduced for the first time to powers of powers. They should use their understanding of what is meant by a power and what they have learnt in the previous chapters to explore what happens when a power is raised to another power. Students often want to move to a procedural method quite quickly, but they should be encouraged to show working and explain their reasoning throughout to ensure that they understand what they are doing rather than just knowing how to work out the answer. Students will continue to use the addition and subtraction laws for indices. Again, they should not be too procedural and they should be encouraged to explain their thinking. By this stage, students have learnt several "rules", and if they fail to show their working and explain their thinking, then they are more likely to confuse them.

Things to look out for

- Adding the powers rather than multiplying
- Multiplying the coefficient by the power rather than raising it to the power
- Not applying a power to a coefficient
- When the base is a number, raising that number to the power rather than leaving the base unchanged

Interleaving opportunities

- Four operations
- Algebraic notation
- Solving equations
- Expanding brackets
- Multiplying and dividing fractions

Models and representations

- Writing calculations in expanded form continues to aid understanding in this chapter. If you know that $a^3 = a \times a \times a$, then you also know that $(ab)^3 = ab \times ab \times ab$, $(17ab)^3 = 17ab \times 17ab \times 17ab$ and $(125a^4b^2)^3 = 125a^4b^2 \times 125a^4b^2 \times 125a^4b^2$.

Behind the questions

Are you ready?

These four questions cover the essential knowledge for this chapter. Question 1 prompts students to remember what is actually meant by a power, as this will be built upon in this chapter. Questions 2 to 4 link back to learning from the previous two chapters on the addition and subtraction laws for indices.

Practice 9.4A

1 requires students to write the calculations out in full. This ensures that they know what is meant by a power of a power and realise that there isn't actually any new learning involved. Students are already familiar with raising a number or expression to a power and with the addition law for indices. Writing this additional step in each part means that students don't need to be explicitly taught anything new.

2 gives students further practice at simplifying powers of powers whilst being encouraged to spot patterns. Students should write the expressions out in full and make connections between the changes made in each part of the question and the effect this has on the answers. By this stage, students should notice that the power of the answer is the product of the powers in the question and that the base remains unchanged. It is important that they can explain why this happens as well as being able to use the result.

3 Here students practise simplifying expressions involving powers of powers. Watch out for them trying to calculate the number to the given power when the base is a number. If this happens, students should be encouraged to write the calculation out in full and compare it with one where the base is algebraic to realise why this is incorrect. Students may think that part l is hard, but they should be prompted to recognise that it is no different from the expressions they have already simplified.

4 highlights the commutative property of multiplication.

5 Here students find missing powers. Watch out for them using trial and improvement. Instead, encourage them to form and solve missing-number calculations to support their thinking.

6 introduces students to powers of powers where there is a coefficient. A common mistake is to multiply the coefficient by the power, which is what Faith has done. Some students may say that she is wrong because 2 is part of the base and so should stay as 2. Instead, students should be encouraged to write the calculation in expanded form to realise that the coefficient needs to be raised to the power.

7 allows students to practise what they have just looked at in question 6. Watch out for students making any of the common mistakes highlighted above.

8 combines all the laws of indices covered so far. Students should be able to explain each step in their working to ensure that they understand each law as well as just being able to find the answer.

What do you think? 9.4A

These three questions require students to think deeply about finding powers of powers. They could be stretched in question 1 to consider how their answers would change if the condition that x and y are positive integers wasn't given. In question 2, watch out for students finding any pair of numbers that have a product of 30 and not considering the base.

Consolidate – do you need more?

These questions revisit the key points covered in the previous exercise, and provide further practice at finding powers of powers. Students are encouraged to write each calculation in expanded form to ensure that they understand what each question is asking. They are then asked to simplify the expressions using their expanded form to support their thinking. If more practice is needed, students could be asked to work in pairs to produce similar questions and challenge each other.

Stretch – can you deepen your learning?

1 generalises what has been covered earlier in the chapter. Students could be prompted to choose values of a, b and c to test their answers and should be able to explain why they don't need to choose a value for A.

2 Watch out for students not multiplying both powers in the bracket by the power outside the bracket. Also, as the expressions become more complex, students might make arithmetic mistakes when raising the coefficient to the power in part b iv and part b v.

3 looks at powers of powers in relation to numbers. You could encourage students to check their final answers using a calculator.

4 and **5** look at powers of powers. Question 4 provides the opportunity to recap multiplying and dividing fractions, whilst question 5 allows students to consolidate their understanding of expanding brackets.

Reflect

Students should explain their learning in their own words as this will clarify their thinking.

10.1 Fractions, decimals and percentages revisited

Small steps

- Convert fluently between key fractions, decimals and percentages (R)
- Calculate key fractions, decimals and percentages of an amount without a calculator (R)
- Calculate fractions, decimals and percentages of an amount using calculator methods (R)

Rationale

This chapter reminds students of the key conversions and calculations covered in Book 1. The images and representations should all be familiar. Students are encouraged to recognise and fluently convert between different forms of numbers, recognising when each is most useful. Students may need reminding about the conversions for $\frac{1}{3}$ and $\frac{2}{3}$ as these are sometimes less commonly used.

Calculation both with and without a calculator is revisited, with a strong emphasis on the decimal multiplier method in preparation for the material in the rest of this block, where this is the key method for looking at percentage increases and decreases, and also links to expressing one number as a percentage of another. Some answers may need rounding, which is not formally covered until Block 12, but students should be familiar with this from KS2. You could choose examples where the need for rounding is minimal.

Things to look out for

- Thinking that 7% = 0.7, for example – discuss the meaning of "per cent" and link to division by 100
- Thinking that 0.3 = $\frac{1}{3}$
- Using non-calculator methods when calculator methods are more appropriate – discuss which approach to take and when
- Students find non-integer percentages more difficult than integers to convert to decimals – encourage them to think of this as dividing by 100 and to use place value tables, or even a calculator, if necessary

Interleaving opportunities

- The elements of this chapter are key to the whole of Block 10 and are revisited throughout
- Return to calculations involving percentages in Block 12, when rounding has been explored in depth
- You can use percentage problems when working with area in Block 14

Models and representations

- The bar model is used to illustrate fractions of amounts and the "build up from 10%" method.
- Number lines are very useful for converting between all three forms. You could extend these beyond one whole/100% if desired, or wait until the next chapter when this is explicitly covered.
- You could use place value charts to support students with challenging decimal conversions.
- Using a series of 100 squares to represent one whole (100%) is useful for decimal conversions when the percentage is an integer.

Behind the questions

Are you ready?

The first four questions assess students' recall of multiplication and division by 100 as this will be a key skill for conversions. Practise further with place value charts if needed. The last question checks recall of key facts and some associated misconceptions.

Practice 10.1A

1 This is a great starting point for finding out which facts students can recall. You may wish to use multiples of $\frac{1}{3}$, $\frac{1}{5}$ and $\frac{1}{8}$, as appropriate for your class.

2 Use place value charts and calculators for support if necessary. Although calculators shouldn't be necessary for dividing by 100, they can be used to support students when they need to work out percentages such as 3.6% of £837.

3 and **4** are reminders of simplifying fractions. You may wish to add more practice here, for example question 3 from the "Consolidate" section.

5 and **6** link finding fractions and percentages of an amount. Preferably these should be answered without a calculator.

7 and **8** go back to conversions in the context of ordering. Discuss as a class the preferred strategy, e.g. converting all values to percentages. Students could create their own lists if more practice is needed.

What do you think? 10.1A

1 Encourage multiple approaches, including subtracting 5% from 50%.

2 explores eighths and fortieths as an extension to the more usual conversions.

3 The bar model should provide support; this was covered in Block 8 of Book 1.

4 looks at the pros and cons of finding percentages by starting from 1%.

Practice 10.1B

1 provides fluency practice at using a decimal multiplier with a calculator. You could compare this method with finding $\frac{a}{100} \times b$.

2 and **3** extend to "harder" decimals. Again, you can use a calculator for support.

4 to **6** add in context and/or interpreting a calculator display. Some discussion may be necessary as giving prices in the form £2.5 is becoming increasingly common in cafes and restaurants, although this is not yet mathematically acceptable. Support may also be needed with rounding – use number lines if necessary.

What do you think? 10.1B

1 illustrates that answers are not always obvious and that a calculator approach is far more efficient than a written one in some situations.

2 interleaves pie charts and provides a simple context. Students could explore other similar labels.

Consolidate – do you need more?

This fluency practice could be used for homework or for revisiting and reinforcing this topic at a later date.

Stretch – can you deepen your learning?

1 This is an exploration in advance of more detailed study in the next chapter. You could start by extending the number lines beyond one whole/100%.

2 This useful relationship was also looked at in Book 1. Using $\frac{a}{100} \times b$ and $\frac{b}{100} \times a$ could now provide a proof, and a formal proof is provided in Book 3.

3 deepens understanding by "working backwards". Students could be extended by using other numbers of their/your choice, or using combinations of fractions, decimals and percentages.

Reflect

Encourage students to think about how and why their methods work as well as their procedural efficacy.

10.2 Using multipliers

Small steps

- Percentage decrease with a multiplier
- Convert between decimals and percentages greater than 100%
- Calculate percentage increase and decrease using a multiplier

Rationale

In this chapter, students formally look at increasing and decreasing a number by a percentage. The decimal multiplier approach is emphasised over the "find the percentage and then add/subtract" method, as this is more versatile and will be key to working with "reverse" percentages. Nonetheless, students should be able to think about non-calculator approaches when appropriate.

Questions of the form "Increase/decrease … by …%" have been generally avoided as they serve little purpose if students are unable to interpret when each operation is appropriate. Words such as "commission", "discount", "service charge", etc. may need explicit teaching.

Things to look out for

- Students may need help reading/interpreting the worded questions – asking questions such as "Will the final price be more or less than the starting price?" will provide support
- Students sometimes just find the percentage, rather than increasing or decreasing a quantity by this percentage – this is explored in question 5 in Practice 10.2A, which uses the multiplier method
- Difficulty with conversion of percentages above 100% – use number lines to illustrate that the same relationships hold as for percentages below 100%
- Students may confuse percentages such as 7% and 70% – compare and contrast increasing a value by these percentages
- Relying on "build up" methods when these are not appropriate – encourage the use of multipliers

Interleaving opportunities

- This chapter develops the learning of Chapter 10.1
- Return to calculations involving percentages in Block 12, when rounding has been explored in depth

Models and representations

- Bar models are very useful for illustrating whether an increase or a decrease has taken place so that students can decide the percentages they need to calculate.
- You could work as a class to extend the number line beyond one whole/100% to see the equivalence of percentages greater than 100%. Students should notice that the same relationships of multiplication and division by 100 still hold.

Behind the questions

Are you ready?

The emphasis is on conversion of percentages to decimals to support the use of the decimal multiplier method. Some questions also focus on the essential fact that 100% is the whole, so that students can use this as a starting point for their conversions and calculations later in the chapter.

Practice 10.2A

1 to **4** scaffold the process of finding a percentage decrease through the use of a decimal multiplier. Bar models could be drawn for support and you could hold a class discussion as to the meaning of words like "commission". You could also compare the multiplier method with working out the percentages and subtracting from the original value.

5 looks at the most common error and is a useful discussion point.

6 and **7** move into the abstract without a context. Again, bar models could be used to support the subtractions, and a number line could help with the conversions.

What do you think? 10.2A

1 looks explicitly at the language associated with percentage increase and decrease.

2 looks at more complex decimal conversions. You could compare mental methods, using a number line and using a calculator.

Practice 10.2B

1 to **3** mirror Practice 10.2A and scaffold the process of finding a percentage increase through the use of a decimal multiplier. Bar models could again be drawn for support. You could also compare the multiplier method with working out the percentages and adding to the original value.

4 and **5** also mirror questions 6 and 7 in Practice 10.2A, and move into the abstract without a context. Bar models could be used to support the additions, and a number line could help with the less familiar conversions of numbers over 100%.

6 encourages students to think about the effect of a given multiplier. Be careful to root out the misconception that, for example, 1.18 increases a quantity by 18% so 0.18 decreases it by 18%.

7 includes both an increase and a decrease. You could build on this by getting students to make up questions based on the new vocabulary they have learnt and challenge each other to solve problems for which they need to decide whether to carry out an increase or a decrease.

What do you think? 10.2B

1 consolidates the first "What do you think?" section by looking at the remaining words.

2 A drawing can help to show that the second percentage increase is by a different number.

3 develops this to look at the effect of two different multipliers.

4 is a useful reminder of finding a percentage of an amount and avoiding doing an unnecessary series of calculations.

Consolidate – do you need more?

This provides a mixture of questions involving both percentage increase and decrease so that students need to think carefully about their choice of method. This could be supplemented as suggested in Practice 10.2B.

Stretch – can you deepen your learning?

1 revisits the idea of repeated changes and challenges students by asking them to work backwards to find out what the possible changes could be. Starting by working on a given number (such as 100 or a multiple of 100) is a useful strategy.

2 introduces the idea of multipliers greater than 2.

3 challenges students to think about "reverse" percentages, which will be covered in Chapter 10.5.

Reflect

Encourage students to reflect on multipliers greater than and less than 1 and to consider when non-calculator approaches may be useful.

10.3 Expressing as a percentage

Small steps

- Express one number as a fraction or a percentage of another without a calculator
- Express one number as a fraction or a percentage of another using calculator methods

Rationale

Before expressing one number as a percentage of another, students need to be able to express one number as a fraction of another, so this chapter looks at both, and focuses on converting fractions to percentages. A visual approach is useful to help students make sense of the learning, and to further this aim questions are mainly set in context rather than asking, for example, "Write 46 as a percentage of 75."

Non-calculator methods of converting fractions to percentages are developed by making the denominator 100. Calculator methods rely on going through the decimal stage. You can link this to multipliers and check answers by finding the associated percentage of the relevant number. For non-exact answers, students are requested to round to the nearest 1%, as rounding to the nearest integer should be accessible. More complex rounding is not covered until Block 12.

Things to look out for

- Difficulty with conversions of the form $\frac{x}{200}$, $\frac{y}{300}$, etc. – examples have been explicitly included so that students don't only think that they need to "multiply up"
- Difficulty with fractions greater than 1 – this should be mitigated by the fact that percentages over 100% were discussed in the previous chapter

Interleaving opportunities

- This chapter has close links with converting between fractions, decimals and percentages, as revisited in Chapter 10.1
- The process of conversion to decimals also builds on the use of multipliers in Chapter 10.2
- Ratio is explicitly revisited
- Knowledge of unit conversions from Book 1 is needed – this will be reviewed in Block 12
- Area is interleaved – this is reviewed in Block 14
- Return to calculations involving percentages in Block 12, when rounding has been explored in depth

Models and representations

- A fraction wall can provide support with familiar fractions, and show the relationships between them, for example to convert fifths to tenths before using $\frac{1}{10} = 10\%$.
- 100 squares are useful for visualising percentages and for comparison with 5 by 4 squares, say, by asking questions like "What percentage does each small square represent?"
- Bar models where the whole is 100%.

Behind the questions

Are you ready?

The first two questions should be accessible following the study of Chapters 10.1 and 10.2, and the last question is similar with the added element of a comparison. Question 3 assesses rounding to the nearest integer. If this causes issues, then invest some time discussing this before moving on to the second part of this chapter.

Practice 10.3A

1 The topic is introduced through identifying fractions of shapes (with denominators that are factors of 100) and converting these to percentages.

2 uses denominators that are not factors of 100.

3 and **4** are set in different contexts. Students should find each fraction first and then convert it to a decimal and a percentage.

4 provides practice in context.

5 to **8** develop the theme of the chapter and include some metric unit conversions and denominators greater than 100.

What do you think? 10.3A

1 reviews "known" conversions.

2 looks at fractions in the context of ratio. Bar models are very useful for preventing the misconception that $2 : 3 = \frac{2}{3}$.

3 looks at "*a* as a percentage of *b*" abstractly, including percentages greater than 100%. Diagrams could be a useful support.

Practice 10.3B

1 scaffolds writing a fraction, converting this to a decimal and then to a percentage using a calculator. Students may need support with the rounding so check this before moving on as this is vital for the remainder of the exercise.

2 to **7** remove the scaffolding but follow the same pattern in different contexts. You may wish to model some of these or provide some scaffolding for those who need it.

8 brings in an element of comparison.

9 interleaves area and is a good starting point for further investigation. You could return to this after Block 14 and include areas of more complex shapes.

What do you think? 10.3B

1 brings in an element of comparison.

2 Students need to take care to use the correct numbers. They can either perform the subtraction first or work out (3000 ÷ 17 000) × 100 and then subtract from 100%.

3 As in Practice 10.3A, this looks at "*a* as a percentage of *b*" abstractly but now using a calculator. Percentages greater than 100% are again included. You might wish to have a discussion on estimation here.

Consolidate – do you need more?

Here both calculator and non-calculator methods are revisited with further fluency practice.

Stretch – can you deepen your learning?

1 You could support students by asking them to think of each square as having an area of 100 cm^2. The white and total areas can then be deduced. Part b could look at squares and other shapes.

Reflect

The reflection focuses on when to use known facts and when to use a calculator. You might also want students to reflect on rounding.

10.4 Percentage problems

Small steps

- Work with percentage change
- Choose appropriate methods to solve percentage problems

Rationale

The chapter serves two purposes. Firstly, students extend their learning of writing one number as a percentage of another from Chapter 10.3 to problems where a change has taken place. Students need to think carefully about the denominator of the fraction used to convert to a percentage. The formula percentage change = $\frac{\text{change}}{\text{original}} \times 100$ is used, but, if you prefer, you can write the change as a fraction first and then convert this to a percentage.

Students sometimes get confused about how to approach a percentage problem, so the remainder of the chapter focuses on choice of operation – finding a percentage, increasing/deceasing by a percentage or expressing a number as a percentage. This is a difficult skill, and time needs to be invested in it so that students can make the correct choices.

Things to look out for	Interleaving opportunities
• Using the wrong denominator when finding a percentage change – the use of the formula is helpful here, but you may need to discuss the meaning of the word "original" in this context and how you can determine which number is the original from the question • Unfamiliarity with the language of profit and loss • Difficulty in choosing the correct approach to take with a question, e.g. multiplication or division – this takes practice and patience, and students could work in pairs to discuss a series of questions to identify the way forward and share strategies with the class as a whole before trying some independently	• The first part of the chapter develops Chapter 10.3 • All the learning from Chapters 10.1, 10.2 and 10.3 is needed for students to successfully choose the correct approach in the remainder of the block • Return to calculations involving percentages in Block 12, when rounding has been explored in depth • You can set problems about percentages in almost all other areas of the mathematics curriculum, especially number-based or number-heavy topics

Models and representations

- Use of the bar model to represent a problem is encouraged to help students decide what approach to take. To support students, ask questions such as "Are we looking within the bar, or are we changing the size or the bar?"
- Number lines remain useful for supporting conversions.
- The formula percentage change = $\frac{\text{change}}{\text{original}} \times 100$ is a useful abstract representation to support expressing a change as a percentage.

Behind the questions

Are you ready?

These questions review the conversions and rounding needed to succeed in Practice 10.4A. The other aspects that need review, for example percentage increase and decrease, are explicitly covered within the chapter.

Practice 10.4A

1 and **2** scaffold finding percentage profit and loss.

3 needs careful thought, and students may need help to see that £50 is the original cost – what the shop paid for the trainers – every time.

4 addresses the error of using the final value instead of the original in the formula for percentage change.

5 and **6** look at change in different contexts. Students may need help with the conversion of metric units.

7 removes the scaffolding completely and now requires the use of a calculator.

8 is a multi-step problem.

What do you think? 10.4A

1 looks at an alternative approach. Compare this with the use of decimal multipliers in Chapter 10.2.

2 is a good discussion starter about changes that exceed 100%.

Practice 10.4B

1 demonstrates the two different types of problems that students may confuse. It is a useful starting point for a class discussion.

2 and **3** add context. Drawing diagrams may be useful.

4 The choice of operation should be clear from the table, but look out for students who mix up the additions and subtractions here.

5 to **8** increase the level of interpretation needed. Encourage discussion as students develop their skills.
In question 8, students may need a hint to first find the percentage change in the last 10 years.

What do you think? 10.4B

1 and **2** are simplified versions of tax calculations. You might wish to model one first and encourage clear working.

Consolidate – do you need more?

This reviews the learning of the chapter for further practice or later use.

Stretch – can you deepen your learning?

1 sets up the next chapter. Bar models are useful here.

2 is challenging. You could encourage students to choose values for cost of production and sales and see the effect of their changes.

Reflect

Students could produce a list of questions in context and explain their choice of whether to multiply or divide by referring to these questions.

10.5 Reverse percentages (H)

Small steps

- Find the original amount given a percentage less than 100% (H)
- Find the original amount given a percentage greater than 100% (H)
- Choose appropriate methods to solve complex percentage problems (H)

Rationale

Students explored finding a number given a percentage in Book 1, and this chapter builds on this understanding, in particular to find the original value after a percentage change. Several approaches are used, including working backwards from given facts (e.g. using 30% to find 10% and then 100%), the unitary approach of finding 1% first, and an algebraic approach of forming an equation using a multiplier.

This is a Higher step and should not be covered unless students are already confident with the learning from the earlier chapters in this block, particularly the second part of Chapter 10.4, where students need to choose the correct approach for solving problems. If they are not competent at this, another type of problem is likely to add to their confusion. Reverse percentages are revisited in Book 3 so the topic can be left until then if appropriate.

Things to look out for	Interleaving opportunities
• Mixing up multiplication and division and/or using the wrong multiplier – students need to be encouraged to ask themselves "What percentage do I know? What do I need to find out? What steps can connect these?" Use your chosen examples to model this thinking. • Treating all problems as reverse percentage problems – the second part of the chapter includes all types of questions to help students to identify the correct approach	• Familiarity with the connections between fractions and percentages from Chapter 10.1 is assumed • The use of the multiplier method from Chapters 10.2 and 10.4 is developed • Forming equations from information, as studied in Block 7, is also revisited • The second part of the chapter interleaves Chapter 10.4 as students have to choose the correct approach • Return to calculations involving percentages in Block 12, when rounding has been explored in depth

Models and representations

- A bar model with the percentage and the amounts before and after clearly visible is a good way of representing this type of problem. Careful labelling of the bar model should ensure that the right calculations follow.

Behind the questions

Are you ready?

Understanding the use of decimal multipliers is key to the algebraic approach for solving reverse percentage problems, so this is the focus of these questions.

Practice 10.5A

1 This could be answered using the "known facts" approach, i.e. deducing each answer from the given fact or previous answers, or by the unitary approach. Bar models can support this.

2 Again, dividing to find either 10% or 1% are both valid first steps.

4 to **5** introduce context and numbers that are less likely to be evaluated mentally. Discussion of the algebraic approach may be useful here. Look out for errors such as £4000 × 0.75 in question 5. A bar model showing that 125% is equal to £4000 helps make sense of this.

6 tackles the most likely type of error.

7 to **9** provide further practice with percentages greater than and less than 100%.

What do you think? 10.5A

1 This looks at the result of two changes. Discuss how this could be shown as an equation and how the equation can be solved.

2 further develops the ideas introduced in question 1.

3 involves interpretation and calculation.

Practice 10.5B

1 to **3** are set in simple contexts but need careful consideration about the choice of approach.

4 takes this further. There is one context but this question involves several types of calculations.

5 to **7** provide further practice in context. Drawing a bar model to show what percentage is known supports access to these problems.

What do you think? 10.5B

1 This looks at the result of two changes. Discuss how this could be shown as an equation and how the equation can be solved.

2 explores the connection between multiplying by 1.5 (or $\frac{3}{2}$) and multiplying by its reciprocal, $\frac{2}{3}$.

Consolidate – do you need more?

This further practice exercise involves decision making to identify the correct approach.

Stretch – can you deepen your learning?

1 looks at common errors and misconceptions. This could be used as part of students' reflection on this chapter.

2 This develops the second "What do you think?" section, by looking at repeated changes. Students could start with numerical examples before generalising.

Reflect

The reflection focuses on recognising when a problem is a reverse percentage problem as well as on how to answer these questions.

11.1 Working with large numbers

Small steps

- Investigate positive powers of 10
- Work with numbers greater than 1 in standard form

Rationale

The purpose of this chapter is to be able to write large numbers in standard form. It is essential that students can use and understand positive powers of 10 before they begin work on this new content. Students were introduced to the laws of indices in Block 9, and this knowledge will be transferred to this chapter. Discussion of the rationale behind writing large numbers in standard form may be useful. Students should be able to recognise when a number, $A \times 10^n$, is not written correctly in standard form. They will need to understand the convention that $1 \leqslant A < 10$ and use their understanding of powers of 10 to explore why n must be an integer. Students may be tempted to think that the power of 10 is always the same as the number of zeros, so they should explore why this isn't always the case early in the chapter in order to fully understand how to write a number in standard form.

Things to look out for

- Thinking that the power always corresponds to the number of zeros
- Using a value of A that is outside the correct range
- Ignoring any digits later in a number, e.g. 300 000 000 600 should be written as $3.000\,000\,006 \times 10^{11}$
- When writing a number in ordinary form, just adding zeros rather than multiplying by 10, e.g. writing 3.25×10^4 as 3.250 000

Interleaving opportunities

- Laws of indices
- Multiplying by powers of 10
- Place value

Models and representations

- A place value grid can be used to compare numbers.

10^4	10^3	10^2	10^1	10^0
Ten thousands	Thousands	Hundreds	Tens	Ones
	9	0	0	0
5	0	0	0	0

The headings of the place value grid could be in words or in powers of 10, or both.

Behind the questions

Are you ready?

These three questions provide an opportunity for students to practise the key skills needed in this chapter. Question 1 checks that students can work with basic powers of 10, question 2 involves multiplying by powers of 10 and question 3 looks at multiplying by multiples of powers of 10. When working through question 2, students should recognise that, although in part a they could describe their method as "adding zeros", this quickly falls apart in parts b and c. It is important that students don't describe multiplying by powers of 10 as simply adding zeros. In question 3, students should be encouraged to explain their method as well as calculate the answer: if they think of 20×300 as $2 \times 3 \times 10 \times 100$, it will help them later in the block when multiplying numbers in standard form.

Practice 11.1A

1 recaps learning from Block 9 on powers in relation to powers of 10.

2 Students are asked to write large numbers as powers of 10. Watch out for students being unsure of the place value of 1 million and 1 billion in parts c and f, respectively.

3 Although students are discouraged from simply counting zeros when writing numbers in standard form, the number here is a simple power of 10 so the value given is correct.

4 builds on question 3. Students should start to see why powers of 10 are used to record a number rather than writing it out in full.

5 This is the first time that students have been asked to write numbers in standard form. Watch out for students thinking that because 100 is 10^2, then 700 is 70^2. Instead, they should be encouraged to partition the number into factors and link it to what they've seen previously, for example $700 = 7 \times 100 = 7 \times 10^2$. All of the question parts will give a 1-digit number multiplied by a power of 10 so the focus is on how the power changes in relation to the order of magnitude of the number.

6 Here the numbers will take the form of a decimal number multiplied by a power of 10. Watch out for students writing, for example, 35 000 as 35×10^3. They should be reminded of the correct range of values for *A* when writing a number in standard form.

7 provides the opportunity for students to consolidate their new learning. In part f, students could discuss whether the power of 1 is necessary.

8 Students start to write numbers in ordinary form. Notice that, in set A, the power is the same each time. These questions can be used to explain why the power isn't simply the number of zeros. In set B, notice that the value of *A* is the same in each part, so students can focus on how the different powers of 10 affect the number.

9 gives students the chance to practise converting from standard form to ordinary form. Links should be made back to powers of 10, for example $2 \times 10^4 = 2 \times 10\,000 = 20\,000$.

What do you think? 11.1A

These four questions require students to think more deeply about large numbers in standard form. Question 1 explores the common misconception around adding zeros; now that students have had plenty of practice they should be able to explain why this is incorrect. Question 2 highlights the importance of writing numbers with breaks in them to make them easier to read. Question 3 asks students to consider whether the numbers have been written correctly in standard form, whilst question 4 involves comparing numbers in standard form.

Consolidate – do you need more?

These questions revisit the key points covered in the previous exercise, to ensure that students can convert between ordinary numbers and standard form fluently. If more practice is needed, students could be encouraged to work in pairs to create similar questions for each other to answer.

Stretch – can you deepen your learning?

1 The focus here is on the method rather than the actual answer. Students should be able to explain each step of their thinking.

2 and **3** look at standard form in real-life contexts.

4 and **5** build on the idea covered at the end of the "What do you think?" section by comparing numbers in standard form. Students should use their knowledge of indices from Block 9 to support their thinking.

Reflect

Students should explain their learning in their own words as this will clarify their thinking. They should be able to identify the benefits of working with numbers in standard form, and explain how to do this.

11.2 Working with decimal numbers

Small steps

- Investigate negative powers of 10
- Work with numbers between 0 and 1 in standard form

Rationale

The purpose of this chapter is to be able to write numbers between 0 and 1 in standard form. As a prerequisite to this, it is essential that students can use and understand negative powers of 10. In the previous chapter, students learnt how to write a large number in standard form, and this knowledge will be transferred to this chapter. It may be useful to discuss the rationale behind writing small numbers in standard form. Students should be able to recognise when a number $A \times 10^n$ is and is not written correctly in standard form. They will need to understand the convention that $1 \leqslant A < 10$ and use their understanding of powers of 10 to explore why n must be an integer, and why n being negative doesn't mean that the number itself is negative. Some students may be tempted to think that the power of 10 is equal to the number of zeros, so they should explore why this isn't always the case early in the chapter in order to fully understand how to write a number in standard form.

Things to look out for

- Thinking that the power of 10 always corresponds to the number of zeros
- Using a value of A that is outside the correct range
- Ignoring any digits later in a number, e.g. 0.300 000 000 600 should be written as $3.000\,000\,006 \times 10^{-1}$
- When writing a number in ordinary form, just adding zeros rather than multiplying by 10, e.g. writing 3.25×10^{-4} as 00003.25
- Writing $\div\, 10^n$ rather than $\times\, 10^n$
- Writing the number as negative because the power is negative

Interleaving opportunities

- Laws of indices
- Multiplying by powers of 10
- Dividing by powers of 10
- Place value
- Fraction and decimal equivalents

Models and representations

- A place value grid can be used to compare numbers.

10^0	10^{-1}	10^{-2}	10^{-3}	10^{-4}
Ones	tenths	hundredths	thousandths	ten thousandths
0 •	0	0	1	

The headings of the place value grid could be in words or in powers of 10, or both. Encourage students to see the link between decimal, fractional and standard forms.

Behind the questions

Are you ready?

These three questions provide an opportunity for students to practise the key skills needed in this chapter. Question 1 checks that students can convert fluently between fractional and decimal powers of 10, question 2 involves multiplying by powers of 10 and question 3 looks at working with powers. When working through question 1, students could be encouraged to consider the relationship between the denominator of each fraction and its decimal equivalent. When working through question 2, it is important that students understand the difference between multiplying by powers of 10 greater than 1 and multiplying by powers of 10 less than 1.

Practice 11.2A

1 Students should recognise that each time the power decreases by 1, the answer is divided by 10, and they should be able to explain why this happens. It is also important to notice that although the power may be negative, the answers are all positive. Students could investigate this so that they realise that a positive number raised to any power will always give a positive answer. This will prevent them thinking that a number is negative because its power of 10 is negative.

2 Students can use their work in question 1 to help them with parts a and b, and they should be encouraged to spot patterns between the number and the power to help them with parts c to e.

3 Although students are discouraged from simply counting zeros when writing numbers in standard form, this will give a correct answer when working with numbers between 0 and 1.

4 builds on question 3. By now students should be able to see the purpose of using powers of 10 to record a number rather than writing it out in full.

5 to **9** Students consolidate writing numbers between 0 and 1 in standard form. They are encouraged to spot patterns and make connections along the way. In question 6, students will obtain a non-integer value for *A* for the first time. They could be encouraged to consider the similarities and differences between the question parts. In question 7, students should understand why there is an integer value of *A* in part a i but not in part a ii or a iii.

10 Here students write numbers in ordinary form. In set A, notice that the value of *A* is the same in each part so that students can focus on how the different powers of 10 affect the number. In set B, notice that the power is the same each time. These questions can be used to illustrate how a different number of decimal places changes the answer.

11 gives students chance to practise converting between standard form and ordinary form. Links should be made back to negative powers of 10, for example $2.4 \times 10^{-3} = 2.4 \times 0.001 = 0.0024$

What do you think? 11.2A

These four questions require students to think more deeply about writing small numbers in standard form. In question 1, they should recognise that the 6 will have the same place value in each number when written in ordinary form because the power of 10 is the same. Question 2 prompts students to recognise when a number will have a positive power of 10 and when the power will be negative. Question 3 asks students to consider whether the numbers have been written correctly in standard form, whilst question 4 involves comparing numbers in standard form.

Consolidate – do you need more?

These questions revisit the key points covered in the previous exercise, to ensure that students can convert between ordinary numbers and standard form fluently. If more practice is needed, students could be encouraged to work in pairs to create similar questions for each other to answer.

Stretch – can you deepen your learning?

1 Students should be encouraged to explain their method in detail here rather than just producing the answers.

2 looks at numbers that are not correctly written in standard form. Students should be able to explain why they are incorrect and then adjust them so that they are correct.

3 builds on the idea covered at the end of the "What do you think?" section by comparing numbers in standard form. Students should use their knowledge of indices from Block 9.

Reflect

Students should explain their learning in their own words as this will clarify their thinking. They should be able to identify the benefits of working with numbers in standard form, and explain how to do this.

11.3 Standard form: mental methods

Small steps

- Compare and order numbers in standard form
- Mentally calculate with numbers in standard form

Rationale

The purpose of this chapter is to compare and order numbers written in standard form and then calculate mentally with numbers in standard form. When comparing and ordering, students should recognise that they do not need to convert to ordinary form. Instead, they can compare the powers of 10 or, if two numbers have the same power of 10, then they can compare the values of A. Students could be introduced to this by being asked "Which is greater, 1 hundred or 1 thousand?" They should recognise that although both of these numbers start with 1, it is the place value of this digit that affects the size of the number. This idea can be transferred to comparing numbers in standard form.

Things to look out for

- Only comparing the number and not looking at the power
- Incorrectly comparing directed numbers
- Not recognising that numbers like 13×10^5 are not in standard form
- Multiplying/dividing the powers when multiplying/dividing by an integer

Interleaving opportunities

- Laws of indices
- Multiplying by powers of 10
- Dividing by powers of 10
- Place value
- Algebraic notation
- Comparing and ordering

Models and representations

- A place value grid can be used to compare numbers.

Millions			Thousands			Ones		
H	T	O	H	T	O	H	T	O
	2	6	9	1	1	1	0	3

Behind the questions

Are you ready?

These four questions provide an opportunity for students to practise the key skills needed in this chapter. Questions 1 and 2 are designed to check that students can convert between ordinary numbers and standard form. Question 3 recaps comparing numbers and use of the inequality symbols. When comparing numbers, students should be encouraged to explain their method by stating which digit they focused on and why. In question 4, students are required to work out the answers using an appropriate mental method. As much focus should be placed on applying the method correctly as on obtaining the correct final answer.

Practice 11.3A

1 explicitly looks at two methods for comparing numbers in standard form. Students first convert numbers into ordinary form and then they use these answers to help them to compare the numbers in standard form.

2 and **3** require students to identify numbers that satisfy a given criterion. To answer question 2, students might first focus on finding the numbers that have a power of 10 greater than 5 and those that have a power of 10 less than 5, because they know that when the power is greater the number is definitely greater, and when the power is smaller the number is smaller. They can then compare the values of A in the numbers where the power of 10 is equal to 5. A similar approach can be used in question 3. Here, students could be prompted to explore two different ways of comparing 30 000 with the other numbers.

4 Students could be encouraged to write 1 million in standard form to support their reasoning here. They could establish that any number $A \times 10^n$ in standard form is greater than or equal to 1 million when n is greater than or equal to 6.

7 Watch out for students making incorrect comparisons with directed numbers.

8 Students may choose to write each number either in standard form or in ordinary form to help make the comparison. They should be encouraged to explain their thinking at each step.

What do you think? 11.3A

These questions require students to think more deeply about comparing numbers in standard form. Questions 1 and 2 explore common misconceptions whilst question 3 looks at constraints on a number and is designed to make students think carefully about how numbers can be compared by looking at their powers of 10.

Practice 11.3B

1 looks at multiplying by powers of 10. Focus should be placed on the power of 10 in the final answer and how this relates to the numbers given in the question.

2 ensures that students can correctly identify when a number is or isn't written in standard form. They should be able to articulate their reasoning for each number and could be stretched to predict how the power of 10 will change if the number is rewritten in correct standard form.

3 combines the skills covered in questions 1 and 2. This provides an opportunity for students to consolidate their thinking. They should be encouraged to write down each step in their working.

4 looks at multiplication and division of a number written in standard form by an integer. Students could write the calculations in ordinary form, if necessary.

What do you think? 11.3B

These two questions require students to think more deeply about calculating with standard form. Question 1 looks at when it is or isn't appropriate to work mentally, whilst question 2 looks at finding pairs of values that give a desired outcome.

Consolidate – do you need more?

These questions revisit the key points covered in the previous exercises to ensure that students can compare and calculate with numbers written in standard form. If more practice is needed, students could be encouraged to work in pairs to create similar questions for each other to answer.

Stretch – can you deepen your learning?

1 and **2** look at generalising results. These questions prompt students to think about what they do and don't need to know in order to compare numbers.

3 and **4** look at different values that can be used to complete a statement. In question 3, only one possible answer is required for each part, whilst question 4 formally explores the range of values which satisfy each inequality.

Reflect

Students should explain their learning in their own words as this will clarify their thinking. They should be able to explain how to compare and calculate with numbers in standard form.

11.4 Standard form: with and without a calculator

Small steps

- Add and subtract numbers in standard form
- Multiply and divide numbers in standard form
- Use a calculator to work with numbers in standard form

Rationale

The purpose of this chapter is to ensure that students are confident when calculating with standard form. The idea of "unitising" could be used for support. For example, when calculating 5 eggs plus 3 eggs, eggs is the "unit" and remains unchanged so the answer is 8 eggs. Similarly, when adding 5 elevenths and 3 elevenths, elevenths is the unit, so the answer is 8 elevenths. This idea can be transferred to calculating with standard form: 5 thousands plus 3 thousands is 8 thousands, so $5 \times 10^3 + 3 \times 10^3 = 8 \times 10^3$. Make sure that students give their final answers in correct standard form. For example, if they are working out $5 \times 10^3 + 6 \times 10^3$ they may give the answer as 11×10^3. But 11 is greater than 10, so this isn't in standard form and needs to be adjusted so the value of A is in the correct range, i.e. $11 \times 10^3 = 1.1 \times 10 \times 10^3 = 1.1 \times 10^4$. It is also important that students are able to use a calculator to work with numbers in standard form and to check their answers.

Things to look out for

- Adding/subtracting the powers when adding/subtracting in standard form
- Multiplying/dividing the powers when multiplying/dividing by an integer
- Not giving the answer in standard form, e.g. leaving it with $A \geqslant 10$ or $A < 1$
- Not recognising when two numbers have different powers and adding/subtracting them anyway
- Incorrect use of the index laws when multiplying/dividing with standard form
- Difficulty with entering numbers in standard form into a calculator

Interleaving opportunities

- Laws of indices
- Multiplying by powers of 10
- Dividing by powers of 10
- Place value
- Algebraic notation
- Four operations
- Area and perimeter
- Calculator skills

Models and representations

- Ensure that students are confident using their calculator by explicitly teaching these skills and giving students time to explore the buttons they need to press.

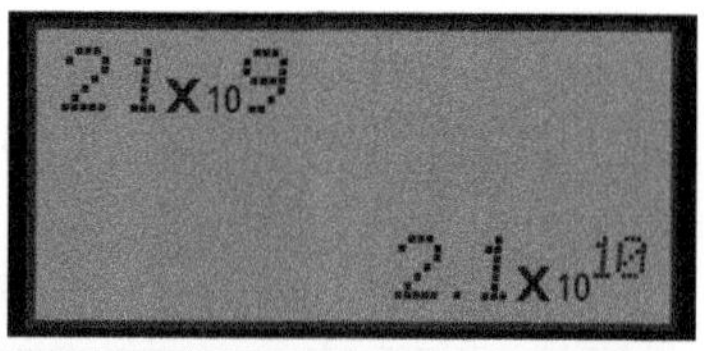

Behind the questions

Are you ready?

These questions provide an opportunity for students to practise the key skills needed in this chapter. Questions 1 and 2 are designed to check that students can convert between ordinary form and standard form using both large and small numbers. Question 3 recaps learning on the laws of indices, with a focus on powers of 10.

Practice 11.4A

1 and **2** focus on rewriting a number so that it is in standard form. This skill will be used throughout the exercise. It is important that students are able to explain why each number is not currently in standard form, as well as using a correct method to adjust each number so that it is. Students should be encouraged to write down and explain their strategy, for example $12 \times 10^4 = 1.2 \times 10 \times 10^4 = 1.2 \times 10^5$.

3 Here students focus specifically on the method rather than the answer. They are encouraged to describe the similarities and differences between the methods, as well as their pros and cons. Although students may have a preferred method, it is important that they understand both whilst being aware of their limitations.

4 and **5** allow students to consolidate their understanding of adding numbers in standard form. They should be encouraged to show their working and reflect on their choice of method. Students should be able to use a calculator to check their answers.

6 Here students are able to practise subtracting numbers given in standard form using their choice of method.

7 to **9** introduce calculations involving standard form in different contexts. In question 7, students calculate the perimeter of a rectangle. They may choose to add the lengths of all four sides or add two sides and then double the answer. Watch out for students not giving their final answer in standard form.

What do you think? 11.4A

These four questions require students to think more deeply about addition and subtraction of numbers in standard form. Question 1 looks at some common mistakes; for example, Abdullah has added 5 zeros to 2.5 rather than multiplying by 10 five times. Question 2 looks at different methods and asks students to consider which is the most appropriate for each question. Question 3 looks at when it is inefficient to calculate in ordinary form and question 4 encourages students to think about when it is possible to get a negative result when calculating with standard form. In question 4, watch out for incorrect explanations such as "$4 - 7 = -3$ so the power will be negative": it is important that students do not link a negative power to a negative number.

Practice 11.4B

1 Students need to use their learning from Block 9 on the laws of indices. Writing $3 \times 10^2 \times 2 \times 10^3$ as $3 \times 2 \times 10^2 \times 10^3$ will help students to find the correct answer. Notice the use of 2 and 3 in part a: this question is designed to highlight the reason why the two values of A should be multiplied whilst the two values of n should be added. Students should be able to explain why this is the case.

2 provides an example where the answer is not in standard form and students need to correct it.

3 builds on question 2 and gives students an opportunity to consolidate their understanding of multiplying numbers given in standard form without using a calculator. They are reminded to give their final answers in correct standard form.

4 moves on to look at division. Students could be encouraged to write each calculation in expanded form.

6 Here a division calculation is broken down to support students. It first gives an answer that isn't in standard form, and then adjusts it accordingly. Part c requires students to use one answer to find another. They should reflect on what's the same and what's different about the questions to support their understanding.

What do you think? 11.4B

These two questions require students to think more deeply about multiplying and dividing with numbers in standard form, by exploring methods of converting to correct standard form and finding pairs of values that produce a given outcome. In question 2, students should be able to explain how the different positioning of x and y affects the answer.

Consolidate – do you need more?

These questions revisit the key points covered in the previous exercises to ensure that students can calculate efficiently with numbers in standard form. If more practice is needed, students could be encouraged to work in pairs to create similar questions for each other to answer.

Stretch – can you deepen your learning?

1 Students could use bar models for support, if necessary. Some students may not recognise that the units on the plank of wood and the units of the pieces cut are different and that one needs to be converted. Students could work through the question twice, once in mm and once in m, to reflect on the similarities and differences.

2 Students should be encouraged to make predictions rather than just trying to work out every possible combination, and they should be able to explain their predictions before checking them.

3 Some students may use the wrong order of operations when applying the formula for the area of a trapezium.

Reflect

Students should explain their learning in their own words as this will clarify their thinking. They should be able to explain how to calculate with numbers in standard form, and explore different methods of calculating.

11.5 Negative and fractional powers (H)

Small steps

- Understand and use negative indices (H)
- Understand and use fractional indices (H)

Rationale

This chapter covers the Higher small steps from the scheme of learning on fractional and negative indices. This will be the first time that students have seen fractional and negative indices other than powers of 10. They should use their learning from earlier in this block and from Block 9 on the laws of indices and negative powers of 10 to support their understanding. Pattern spotting will allow students to use what they already know as a foundation for their new knowledge. Students should already be familiar with the idea of a reciprocal from their work on fractions, and this terminology should be used throughout when working with negative indices.

Things to look out for	Interleaving opportunities
• Multiplying a number by its power rather than raising the number to the power • Thinking that a negative power means that the answer will be negative • Dividing by 2, 3, etc. rather than finding the square or cube root	• Laws of indices • Multiplying by powers of 10 • Dividing by powers of 10 • Algebraic notation • Reciprocals • Powers and roots

Models and representations

- Ensure students are familiar with all of these buttons on their calculator.

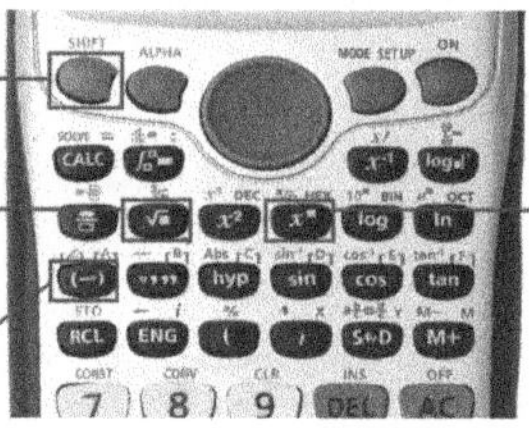

Behind the questions

Are you ready?

Questions 1 and 2 are designed to check that students can raise a number to a given power, both with and without a calculator. Question 3 ensures they can write a number as a power of a given number. Watch out here for students confusing the wording and raising each number to the power of 2, rather than expressing them as powers of 2: it is important that students understand the effect of the slight difference in wording. Question 4 and 5 recap prior learning on powers and roots.

Practice 11.5A

1 reinforces the idea that the power of –1 represents a reciprocal. Students should be able to match each number to its reciprocal.

2 looks at negative powers other than –1, specifically applied to the number 2. Students should spot patterns between rows in the table and should be able to make predictions about other powers not in the table; for example, if 2^{-5} is $\frac{1}{2^5}$ then how would 2^{-99} be calculated?

3 explores the common misconception that a negative power gives a negative answer. Students should recognise straight away that this is incorrect and be able to use the concept of a reciprocal to explain why.

4 and **5** Here students have the chance to consolidate their understanding of working with negative indices. Question 4 is scaffolded whilst question 5 is not. Students should be encouraged to replicate the working exemplified in question 4 to support them with question 5.

6 Here students look back at their learning on order of operations to explain why these are different.

7 Try to prevent students going straight to the decimal answer as they are likely to make mistakes. Instead, encourage them to always write each answer as a fraction, before using their knowledge of fraction and decimal equivalents to convert it to a decimal.

8 Students should notice that the signs of the powers are different in each pair and they could be encouraged to explore why. They could go on to consider questions such as "If $x = a^b$, then what is $\frac{1}{x}$ as a power of a?"

What do you think? 11.5A

So far students have seen negative indices applied to integers, but in question 3 students will apply the fact that the power of –1 represents a reciprocal to fractions. Students should be encouraged to make predictions before checking their answers on a calculator.

Practice 11.5B

1 Students should recognise that the power of one half means finding the square root and should be able to explain why, for example, $16^{1/2} \times 16^{1/2} = 16$ and $4 \times 4 = 16$ so $16^{1/2} = 4$.

2 exposes students to the common mistake of just finding a fraction of the amount instead of a root. Students could explore why this is incorrect by cubing 9, before going on to calculate the correct answer.

3 gives students the opportunity to practise working with fractional powers. Watch out for students overcomplicating parts j and k just because the number raised to the power is negative. Students could be stretched to consider why it is not possible to calculate the value of a negative number raised to any fractional power with an even denominator, e.g. –125 to the power of one half.

4 Students could work backwards to help them to answer this question. They should be able to explain how they know that each pair is matched correctly.

5 Here students need to find the missing power. They will need to think of the connection between the number raised to the power and the answer; for example, $7^2 = 49$, so to find 7 as a power of 49 they will need to use the reciprocal of 2, i.e. $7 = 49^{1/2}$.

What do you think? 11.5B

In question 1, they are asked to explain why the power of one half is the same as finding the square root. Question 2 looks at why it is not possible to raise a negative number to a fractional power if the denominator of the power is even, and question 3 builds on this to consider which calculations can be worked out and which cannot.

Consolidate – do you need more?

These questions revisit the key points covered in the previous exercises to ensure that students can calculate with fractional and negative indices. If more practice is needed, students could be encouraged to work in pairs to create similar questions for each other to answer.

Stretch – can you deepen your learning?

1 Students should be encouraged to break each calculation down into more manageable parts rather than calculate it one go; for example, in part a they know that $3^2 = 9$ and that $2^{-1} = \frac{1}{2}$, so they can find the answer by working out 9 multiplied by one half.

2 Students use their knowledge of fraction and decimal equivalents to explain why Emily is correct.

3 Students build on what was explored in question 2 and use fraction and decimal equivalents to find decimal powers.

4 and **5** combine knowledge of fractional and negative indices. Again, students should be encouraged to break the question down into more manageable parts using their knowledge of powers of powers; for example, $16^{-1/2} = (16^{1/2})^{-1} = 4^{-1} = \frac{1}{4}$

Reflect

Students should explain their learning in their own words as this will clarify their thinking. They should be able to explain how to work with fractional and negative indices and use examples to support their reasoning.

12.1 Rounding, estimating and calculating

Small steps

- Round numbers to powers of 10, and to 1 significant figure (R)
- Estimate the answer to a calculation
- Calculate using the order of operations (R)

Rationale

Students have needed to round numbers in many chapters of this book already. This chapter provides an opportunity to look again in depth at the learning from Book 1, in particular rounding to the nearest 1, 10, 100, 1000 and to 1 significant figure. The latter is used as a basis for estimating the answer to a calculation. You can also use this chapter to extend students' knowledge of significant figures by including rounding to 2 or more significant figures if desired. Dividing by a decimal is explored to support estimation of answers to calculations of the form $\frac{a}{b}$, where $0 < b < 1$.

The correct order of operations is another important aspect of the curriculum that needs revisiting, and this has also been included here, both with simple numbers and when performing estimations.

Things to look out for

- Confusing significant figures and decimal places – decimal places will be covered separately in the next chapter
- Not including enough zeros, e.g. writing the answer as 6 when rounding 617 to 1 significant figure
- Rules like "BIDMAS" might be misinterpreted as implying that addition comes before subtraction, for example – emphasise that multiplication and division, and addition and subtraction, have equal priority
- Thinking that dividing by a number less than 1, e.g. 0.5, makes a number smaller
- Misuse of the equals sign, e.g. writing "697 = 700" instead of "697 = 700 to the nearest hundred"

Interleaving opportunities

- In the next chapter, decimal places are explored and should be contrasted with significant figures
- The formula for the circumference of a circle is revisited
- Converting currencies is revisited
- Rounding will be needed again when the area of a circle is explored in Block 14
- You could revisit finding a percentage of an amount with a decimal multiplier and/or expressing one number as a percentage of another, using rounding to give the answer to a desired number of significant figures

Models and representations

- The number line is an ideal model for illustrating rounding to the nearest 10, 100, etc. as students can see which multiple the number is physically nearer to. Students may need help with identifying the midpoints if using this model for rounding to 1 or more significant figures with end points such as 0.03 and 0.04.
- Counters can be used for calculations such as $3 + 4 \times 2$ and $3 \times 4 + 2$ to model the order of operations.

Behind the questions

Are you ready?

The first three questions assess students' expected knowledge from Book 1. If problems arise, these are covered again in the main text and will not need separate intervention. The last question is a reminder about circumference, as this is one of the contexts used in the chapter.

Practice 12.1A

1 to **3** are fluency practice at simple rounding that can be supported by the use of number lines.

4 uses numbers rounded to 1 significant figure to find estimates. This might need modelling, but the numbers have been chosen to make the resulting calculations easy.

5 and **6** revisit the formula for the circumference of a circle.

7 and **8** revisit currency conversion and area, respectively. You could compare the estimates with the exact answers.

What do you think? 12.1A

1 looks at estimating square roots. Students may need support to look at nearby square numbers to justify their conclusions.

2 extends to rounding to 2 or more significant figures. You could use the numbers from questions 1 to 3 in Practice 12.1A for further practice if required.

3 is well worth a class discussion. Although rounding to 1 significant figure is the most common way of finding estimates, sometimes other rounding is more appropriate, as in this case.

Practice 12.1B

1 provides practice of using expected knowledge.

2 uses equivalent fractions to support understanding of how to divide by a decimal.

3 uses the learning from question 2 for working out other estimates. This is explored in context in question 4.

5 links order of operations and estimation.

What do you think? 12.1B

1 reminds students that not all decimals need rounding and that exact fractional equivalents can be used. They may also need to be reminded that $\div \frac{1}{4}$ is the same as $\times$ 4.

2 interleaves expressing a number as a percentage. You could provide further practice here.

3 extends to more complex calculations. If you challenge students to create their own, a discussion of what makes a "good" question is useful; for example, if rounding results in a denominator such as 7 that is not a factor of the numerator, then the calculation is still complex and the purpose of estimation has not been achieved.

Consolidate – do you need more?

This provides fluency practice in the key learning from the chapter.

Stretch – can you deepen your learning?

1 explores physical estimation and can be used for class discussion alongside question 4.

2 is a development of question 1 and indicates how using a familiar object can help to promote students' understanding.

3 The square root algorithm can be found by an internet search.

Reflect

The reflection has been chosen to include discussion of division by decimals.

12.2 Decimal places

Small steps

- Round numbers to a given number of decimal places
- Understand and use error interval notation (H)

Rationale

This formal look at rounding to decimal places will build on students' previous knowledge of this from KS2, and it links to their study of significant figures and powers of 10 in Book 1 and the previous chapter. Using number lines is again the recommended approach.

Number lines are also recommended for reversing the process in the Higher step of giving an error interval for a rounded number. Students will need to be confident in the use of inequality symbols.

Things to look out for

- Confusing significant figures and decimal places – compare with the learning in the previous chapter
- Rounding 6.97 to 6.10 ("six point ten"), for example – use number lines and encourage students to think of the two possible values that the number to be rounded lies between
- Difficulty in finding the midpoint – discuss writing 0.7 and 0.8 as 0.70 and 0.80, for example, to help, and repeat for numbers with 2 or 3 decimal places
- Difficulty in reading/writing error intervals – this will need careful modelling

Interleaving opportunities

- This chapter builds on Chapter 12.1
- Calculator skills are developed
- Currency conversions are revisited
- Inequality symbols were introduced in Block 7
- Any future rounding, e.g. of areas in Block 14, can practise significant figures or decimal places as desired – note that rounding to a number of significant figures is generally more useful as it can be applied to numbers of any size whereas rounding to a number of decimal places is limited in its usefulness

Models and representations

- Place value grids will support the understanding of the relative sizes of numbers.
- Use number lines to establish "to the nearest" as in Chapter 12.1.

Behind the questions

Are you ready?

Question 1 revisits place value, which is essential for progress in this chapter. If students can round to the nearest integer, as in question 2, confidently, this bodes well for extending this skill to decimal places. They also need to be able to read decimals correctly, as in question 3. Saying, "four point twenty-two", for example, is likely to lead to error.

Practice 12.2A

1 A number line is provided to support students' rounding.

2 Students now use their own lines, using question 1 as a model.

3 challenges common misconceptions.

4 and **5** repeat questions 1 and 2 but extend to the second decimal place. Help students to see that the process is identical.

6 to **8** provide contexts for students' rounding: using a calculator, circles and currency conversions. All of these may need some reminders before students tackle the questions.

What do you think? 12.2A

1 checks understanding of place value.

2 compares significant figures and decimal places.

3 looks at the accuracy of an estimate.

All of these questions are starting points for wider discussions.

Practice 12.2B

1 looks at what values might round to a given integer as an introduction to the idea of an error interval.

2 reverses this by starting with the interval.

3 extends to error intervals with decimal places. Number lines could be useful here.

4 Number lines again could be useful for contrasting significant figures with decimal places.

5 needs care with units.

6 would benefit from drawing a number line for each part.

7 looks at significant figures and contrasts discrete and continuous variables. It might be worth revisiting this vocabulary.

8 also looks at significant figures.

What do you think? 12.2B

1 This is the basis for an informal discussion of upper and lower bounds, which students will study at KS4. Encourage students to think about all combinations of both values.

2 challenges understanding by looking at error intervals of different widths.

Consolidate – do you need more?

This provides fluency practice on decimal places, including some in the contexts explored earlier.

Stretch – can you deepen your learning?

1 examines a different notation.

2 builds on question 1 in the second "What do you think?" section to look at division. This could be extended to look at other contexts and/or the abstract maximum/minimum values of $a + b$, $a - b$, ab and $\frac{a}{b}$ for rounded a and b.

3 looks at when rounding might occur in other contexts.

Reflect

Students reflect explicitly on the difference between significant figures and decimal places and, if appropriate, error intervals.

12.3 Calculating in context

Small steps
- Convert metric measures of length
- Convert metric units of mass and capacity
- Calculate with money
- Solve problems involving time and the calendar

Rationale

This chapter is designed to help students practise their number skills in a wide variety of contexts. They revise metric units from Book 1 and look at money in increasingly complex problems. Finally, problems with time and the calendar are explored; sometimes, lack of engagement with problems of this kind means that students may be unaware of some basic facts that they need to know.

Things to look out for	Interleaving opportunities
• Mixing up the meanings of metric prefixes, especially "milli-" – use all contexts (mass, length and capacity) to reinforce the meanings • Not knowing the number of days in each month, or in a year/leap year • Difficulty in adding/subtracting times on a calculator	• Revision of multiplication and division by powers of 10 • Using non-calculator skills • Forming and solving equations • Confidence at working with metric units is needed before exploring the Higher step of converting between units of area and volume in the next chapter

Models and representations
- Place value grids support multiplication and division by 10, 100 and 1000.
- Place value counters are an alternative.
- Provide actual calendars for students to explore, or use calendar functions on phones instead.
- Use empty number lines to help with time calculations.

Behind the questions

Are you ready?

These questions will give a sense of what unit conversions students have remembered from Book 1, and of their confidence with multiplication and division by powers of 10. Any difficulties that arise will be addressed as students work through the chapter.

Practice 12.3A

1 Students need to be able to use the relationships between cm, mm and m for this question. The diagram before Example 1 may be useful.

2 explores different ways of solving a problem with different units.

3 and **4** look at mass and bring in the prefix "kilo-". You might wish to relate this to kilometres.

4 provides practice in context.

5 to **8** look at capacity.

What do you think? 12.3A

1 and **2** provide an element of challenge by looking at units in the context of interpreting charts. Students could find other similar charts online to create more questions or explore further.

Practice 12.3B

1 and **2** concentrate on converting pounds and pence in context, and address a common misconception.

3 to **6** bring in an element of calculation, which is intended to be carried out without a calculator.

7 could be approached with or without a calculator.

What do you think? 12.3B

1 and **3** are linked to forming and solving equations.

2 interleaves ratio.

Practice 12.3C

1 and **2** are fluency questions using hours, days, weeks and months.

3 brings in calculations with minutes.

4 provides practice in context.

5 to **7** are time calculations that would benefit from the use of a number line.

8 Students may need to be reminded about standard form.

What do you think? 12.3C

1 and **2** These questions explore students' knowledge of the length of a year, with question 2 requiring knowledge of how to identify a leap year.

3 will benefit from the use of a number line.

Consolidate – do you need more?

It could be very useful to return to this exercise some weeks after completing the chapter, particularly if students need a lot of assistance/reminders first time.

Stretch – can you deepen your learning?

1 involves some fraction calculations.

2 is an interesting discussion point. Many students use imperial measures but are unsure of the conversions.

3 could be extended to a general investigation of factorial numbers.

4 reinforces leap years.

Reflect

Students should record any new learning carefully for future reference.

12.4 Units of area and volume (H)

Small steps

- Convert metric units of area (H)
- Convert metric units of volume (H)

Rationale

This Higher chapter looks at conversions between units of area and volume, and challenges common errors such as thinking that 1 cm^2 = 10 mm^2. The approach taken is to compare areas and volumes of the same shapes using different units and generalise to find the appropriate multiplying factor. Volume is not explored in depth until Book 3, but students may remember the concept from primary school, and a reminder of the basic idea is included here to enable simple exploration.

Things to look out for	Interleaving opportunities
• Using linear conversions for area and volume measures – the chapter explores this misconception	• This builds on unit conversion in Chapter 12.3 • Map scales are revisited • Area formulae from Book 1 are assumed • Areas of trapezia and circles are looked at in Block 14, so the conversions can be revisited there

Models and representations

- Squared paper can be used to show areas in both cm^2 and mm^2. Some graph paper uses 1 mm^2 squares and will show directly that 1 cm^2 = 100 mm^2.
- Unit cubes can be used to make cuboids to demonstrate why volume = length × width × height.

Behind the questions

Are you ready?

Finding areas and volumes is the vehicle used to derive general results, so reminders are included here. Similarly, conversion of units of length is included. This Higher step may be omitted if students find converting lengths difficult.

Practice 12.4A

1 and **2** These questions explore using 1 cm^2 = 100 mm^2, both abstractly and in the context of finding areas.

3 moves on to converting between m^2 and cm^2. Check that students convert both sides to 400 cm before working out part b.

4 and **5** then use the m^2 to cm^2 conversion. Discuss when it is appropriate to use multiplication and when to use division.

6 provides questions set in context.

7 extends to looking at m^2 and mm^2. Students can compare using 1 m = 1000 mm with using the two previous results (cm^2 to mm^2, and also m^2 to cm^2) to find the answer.

What do you think? 12.4A

1 involves map scales.

2 brings in km^2.

3 looks at the less familiar measure, the hectare.

Practice 12.4B

1 and **2** follow the pattern of Practice 12.4A, this time looking at volume.

3 moves on to converting between m^3 and cm^3.

4 and **5** then use the m^3 to cm^3 conversion. Discuss when multiplication is appropriate and when to use division.

6 provides questions set in context.

7 extends to looking at m^3 and mm^3. Students can compare using 1 m = 1000 mm. Standard form is recommended for the answer, and students may make, or be guided to, the link that $(10^3)^3 = 10^9$ and look for similar links in other conversions.

What do you think? 12.4B

1 and **2** look at volume problems in context.

3 The fact that 1 litre = 1000 cm^3 is useful here. Students may also be aware that 1 ml of water has a mass of 1 g.

Consolidate – do you need more?

This provides fluency practice for all of the conversions.

Stretch – can you deepen your learning?

1 explores area factors beyond powers of 10. Link to question 2 in the second "What do you think?" section.

2 checks students' familiarity with what these measures "look" like.

3 revisits the connection with litres.

4 explores using the formula for the volume of a cylinder, including working backwards.

Reflect

The reflection challenges students to explain why two common misconceptions are wrong.

13.1 Angles and shapes revisited

Small steps

- Understand and use basic angle rules and notation (R)

Rationale

The aim of this chapter is to revise the angle rules covered by all students in Book 1. The emphasis is on recognising the correct relationships to apply when finding unknown angles. Students should be encouraged to give full mathematical reasons for their answers, rather than numerical ones only.

Mathematical notation, including the use of hatch marks and three-letter notation, is also covered. Students may need a reminder of both of these. Some questions have been designed to revise forming and solving equations.

Things to look out for	Interleaving opportunities
• Incorrect or incomplete language (e.g. "because it's a triangle") when giving reasons – use and expect correct language, and encourage students to challenge you and each other • Not spotting vertically opposite angles when they aren't "vertical" – remind students to look for a vertex formed by two lines crossing each other • Misunderstanding the meaning of "angles on a straight line add up to 180°" – emphasising the meaning as "adjacent angles on a straight line" is recommended • Students may try to measure rather than calculate the angles	• Calculations with basic number • Forming and solving equations is revisited • Ratio is revisited • The rules in this chapter are revisited and built on in the remainder of this block

Models and representations

- Straws can be used to form shapes and illustrate vertically opposite angles.
- Dynamic geometry software can be used to illustrate and remind students of the rules.
- Line segments can be created on geoboards and geoboard apps to demonstrate relationships.

Behind the questions

Are you ready?

These questions assess students' prior knowledge of types of angles, and drawing and measuring angles. Although not needed in this chapter, the aim is to remind students of key relationships where angles sum to either 180° or 360° ahead of the calculations that follow. If these skills need more practice, there are opportunities in Chapters 13.2 and 13.3. Three-letter angle notation is also included and will be revisited throughout this block.

Practice 13.1A

1 revises angles at a point, on a straight line, and vertically opposite angles. Students need to be confident with these before proceeding to the next questions. You may wish to cover these separately.

2 The angle sum in a triangle and in a quadrilateral are included here.

3 Students use reasoning and their knowledge of the sum of adjacent angles on a straight line.

What do you think? 13.1A

1 It is expected that students will form and solve simple equations. They may need to be alerted to this approach.

2 and **3** are designed for students to show their knowledge of angles in quadrilaterals, particularly special quadrilaterals. This is the focus of their learning in Chapter 13.3.

Practice 13.1B

1 Students need to use the properties of equilateral triangles and the fact that "base angles in an isosceles triangle are equal". Discuss the meaning of "base" in this context in your modelling.

2 adds in quadrilaterals and forming and solving equations, which may need revision. Note that part f involves unknowns on both sides. Set this question only if your students have covered this Higher step.

3 Students may need support to set out a clear solution using three-letter angle notation and giving reasons, although the calculation itself is not difficult.

What do you think? 13.1B

These questions give opportunities for students to show their reasoning skills using facts about equilateral and isosceles triangles.

Consolidate – do you need more?

This provides further practice at using the skills covered in the previous exercises. Use your knowledge of students' progress to decide which questions are suitable and which may be too challenging.

Stretch – can you deepen your learning?

1 and **2** provide more opportunities for using reasoning. This will be developed in Chapter 13.4, when students look at interior and exterior angles in detail.

3 revises the meaning of "regular" and targets a common misconception.

4 Students could take this further by exploring other ratios and by using dynamic geometry software to help them with the sketching.

Reflect

This reflection could form the basis of a display.

13.2 Investigating parallel lines

Small steps

- Investigate angles between parallel lines and the transversal
- Identify and calculate with alternate and corresponding angles
- Identify and calculate with co-interior, alternate and corresponding angles

Rationale

Students following the Higher strand will have explored angles in parallel lines in Book 1, but for many students this will be new. The key focus of this chapter is identifying the different types of angle relationships. There is some calculation involved, but more complex problems are left until Chapter 13.5. It may be useful to help students recognise the shapes made by the different types of angle relationships ("F", "Z" and "C"), but you will need to emphasise that correct vocabulary should be used when giving reasons in a solution. Students need to be comfortable with recognising the relationships with pairs of parallel lines in all orientations, not just vertical and horizontal. This learning is reviewed in Book 3.

Things to look out for

- There is a lot of vocabulary, and students may find it hard to remember it all – make associations with the everyday meanings of the words, e.g. alternate angles are on opposite sides of the transversal and "to alternate" means to swap around, etc.
- Students may not realise that the rules work both ways, e.g. that if the co-interior angles add up to 180°, then the two lines must be parallel – you could include examples of using this to decide whether a quadrilateral is a trapezium or not
- Mixing up the rules – invest time on the rules individually before moving on to using all of them together

Interleaving opportunities

- The rules from Chapter 13.1 are needed here
- This learning is built on during the remainder of the block, especially in Chapter 13.3, where parallel lines in quadrilaterals are explored, and in Chapter 13.5, where students look at more complex problems

Models and representations

- Dynamic geometry software is very powerful for illustrating the rules for angles in parallel lines. Be careful to emphasise that demonstration is not the same as proof.
- Manipulatives can be helpful for showing this, for example by moving a transversal straw over a pair of parallel straws.
- Students could construct a pair of parallel lines and a transversal and measure angles as another demonstration, particularly if protractor skills need practising.

Behind the questions

Are you ready?

The meaning of "parallel" as opposed to "perpendicular" is revised. There are also some fluency questions on the skills practised in Chapter 13.1 that will be needed for the simple calculations in this chapter. In particular, students need to be confident with adjacent angles on a straight line and vertically opposite angles.

Practice 13.2A

1 and **2** focus on recognising alternate and corresponding angles, respectively. If students need more practice after this, they could draw their own diagrams and challenge each other. This step is vital before moving on to the later questions.

3 extends the recognition of alternate and corresponding angles by looking for both relationships in the same diagram. This could be supplemented by drawing more diagrams in a variety of orientations.

4 Students now apply their learning to finding missing angles. A class discussion on part b, looking at the many possible approaches, is worthwhile.

5 moves the learning on by looking for the relationships in several different diagrams in different orientations.

6 includes three-letter notation. The length of QP may make the angle relationship more difficult to see. You may wish to discuss this one as a class and ask students to produce similar “different looking” sets of parallel lines and transversals.

7 is a reminder that corresponding angles are only equal if the lines are parallel.

What do you think? 13.2A

These questions are more complex and should be tackled only if students have succeeded in the main exercise. Some of the equations also involve unknowns on both sides.

Practice 13.2B

1 Students learn to recognise co-interior angles. Again, if students need more practice after this, they could draw their own diagrams and challenge each other.

2 and **3** check that students can recognise all the types of angle relationships learned.

4 and **5** provide practice at using the angle relationships. Check students are able to articulate their reasoning using correct language.

What do you think? 13.2B

1 Proof of this kind is developed further in Chapter 13.5. Students could start by drawing a bar model for $a + b = 180°$

2 helps students to recognise co-interior angles even when the line segments are not continued.

3 requires students to use angle facts to establish whether lines are parallel, in contrast to the more usual type of question involving finding missing angles given sets of parallel lines.

Consolidate – do you need more?

Fluency practice is provided. Students could again create their own similar questions to provide further challenge or practice to help them to recognise the different angle relationships.

Stretch – can you deepen your learning?

1 looks at more than one pair of parallel lines in each direction.

2 Students may need a hint to draw a line through B parallel to the given parallel pair. They should then be able to recognise the alternate angles.

3 involves another “truncated diagram”.

Reflect

Recording and discussing their knowledge will help students to cement their learning.

13.3 Angles in quadrilaterals

Small steps

- Construct triangles and special quadrilaterals
- Investigate the properties of special quadrilaterals
- Identify and calculate with sides and angles in special quadrilaterals
- Understand and use the properties of diagonals of quadrilaterals (H)

Rationale

Students are already familiar with the angle sum of a quadrilateral and the names of special quadrilaterals. Here they move on to looking more formally at their properties, and the chapter includes a Higher step where angles and lengths formed by diagonals are explored in detail. This step could be omitted if students are becoming overloaded with too much new information.

This chapter is also used to revisit the skills of construction using a ruler, a pair of compasses and a protractor. As this equipment is not regularly used in the curriculum (outside of this topic), students may need support and guidance to enable them to use it effectively.

Things to look out for

- Difficulty using mathematical equipment – students may need reminding to hold pairs of compasses at the top, not at the arms, when constructing, for example, or how to place and read off a protractor
- There is a lot of information, and students can become frustrated by trying to recall it – emphasise that if they are familiar with the structures of the shapes, then many of the properties can be deduced from a diagram, rather than having to learn a long list of properties
- Students tend to focus on side rather than angle properties – emphasise the relationship between the two, e.g. that if a triangle has a pair of equal sides, then it must also have a pair of equal angles, and vice versa
- To help them to work with the angles in special quadrilaterals, students should be able to use the properties of angles in parallel lines, especially co-interior angles

Interleaving opportunities

- The angle sum of a triangle and a quadrilateral are key facts, alongside the relationships between angles in parallel lines
- The skills of using mathematical equipment are developed in Chapter 13.6
- Chapter 13.6 also links properties of diagonals to the constructions of bisectors
- Shapes will be explored again in Block 15 when students will study their symmetry properties

Models and representations

- Geoboards, geoboard apps or squared paper are useful for comparing quadrilaterals.
- Dynamic geometry software can be used to illustrate properties, including those involving diagonals.
- Set squares are useful for constructing parallel lines.

Behind the questions

Are you ready?

Questions 1 and 4 assess students' recall of the properties of triangles and quadrilaterals, which will be revisited throughout the chapter. Likewise, the use of a ruler and protractor, which is practised in questions 2 and 3, is crucial for this learning.

Practice 13.3A

1 The SSS construction is demonstrated in Example 1.

2 provides a step-by-step reminder of the SAS construction.

3 You may need to model this if students have forgotten or cannot deduce the method from question 2.

4 provides practice at all three types with no scaffolding. You may wish to discuss this question as a class.

5 to **7** use the skills with constructing triangles to construct quadrilaterals. Properties are questioned within these, and you could also hold a summary discussion.

What do you think? 13.3A

Students are guided to use known angle facts to deduce rules about angles in a parallelogram, trapezium and kite. They may need reminding about an isosceles trapezium as some will associate the word "isosceles" with a triangle only.

Practice 13.3B

1 and **2** Students use the properties they have discovered to find missing angles in a parallelogram and a rhombus.

3 and **5** include other angle facts.

4 uses three-letter angle notation.

What do you think? 13.3B

1 requires reasoning – discuss why only one angle is needed.

2 Students need to interpret complex language. Encourage them to think about each word and draw a sketch.

3 Students should be looking for shapes within shapes in questions like this.

Practice 13.3C

1 and **2** Accuracy may be an issue here. Discuss measured answers versus expected answers. You could also compare with measurements from dynamic geometry software.

3 Students could use construction, properties they already know and/or dynamic geometry software to verify their ideas.

4 and **5** Students apply their learning to finding missing angles. Many of the problems are multi-step. Encourage students to look for any results that they can deduce immediately and use these to work towards the required angles.

What do you think? 13.3C

1 and **2** aim to promote a discussion about whether diagonals can be vertical or horizontal. You may wish to show students that one of the diagonals of a delta is external.

3 Students are required to interpret formal statements to check their learning.

Consolidate – do you need more?

More practice is provided for the first two parts of the chapter.

Stretch – can you deepen your learning?

1 Encourage students to explain why the formula works. The (n – 3) term arises because there is no diagonal from a vertex to itself or to its adjacent vertices, and you divide by 2 as each diagonal is counted twice.

2 Congruency is formally covered in Book 3. Here students are using observation rather than the rules of congruency.

Reflect

This is another reflection task that could be used to produce a display.

13.4 Exterior and interior angles

Small steps

- Understand and use the sum of the exterior angles of any polygon
- Calculate and use the sum of the interior angles in any polygon
- Calculate missing interior angles in regular polygons

Rationale

Through exploration, students notice that the exterior angles of any polygon add up to 360° and use this to solve problems. Building on their knowledge of the angle sum of a triangle, students are then led through the process of deducing that the interior angle sum is $(n - 2) \times 180°$. You may also wish to look at joining the vertices to the centre of the polygon and using the equivalent formula $180n - 360$, especially with students who looked at the first formula in the Higher step in Book 1. When working with regular polygons, it is often easier to start with the exterior angles.

Things to look out for	Interleaving opportunities
• Difficulty in finding exterior angles - beware of the misconception of subtracting an angle from a full turn instead of a straight line. Students may also need a reminder to extend the lines in the same direction from all vertices. • Confusion about the meaning of "regular" – emphasise that sides and angles must be equal • Treating all polygons as regular • Misremembering the names of polygons – associate these names with other words, e.g. pentathlon, decathlon, etc.	• The angle sums of a triangle and a quadrilateral are revisited • Other angle rules, especially adjacent angles on a straight line, are used • There are opportunities to form and solve equations • Students could be asked to practise construction by making the polygons in the questions, and to compare the accuracy of their drawings with the answers found by calculation

Models and representations

- Geoboards, geoboard apps and spotty paper are all good forms in which to draw polygons and extend lines for exterior angles and/or split polygons into triangles.
- As shown in the chapter, using a rotating pencil can demonstrate the sum of the exterior angles.
- Dynamic geometry software can be used for both aspects of the chapter.

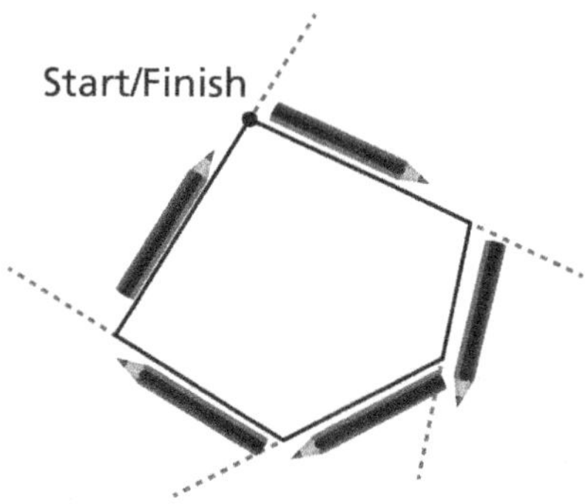

Behind the questions

Are you ready?

The first question tests prerequisite knowledge of the sum of the angles in a triangle and on a straight line. Names of some polygons are included and you could use this as a starting point for a discussion. The last question illustrates that polygons may not take their "expected" form!

Practice 13.4A

1 to **4** use students' previous knowledge to lead to the generalisation that the exterior angles of any polygon sum to 360°.

5 looks further into this conjecture.

6 provides fluency practice at using the fact that the exterior angles of a polygon sum to 360°.

What do you think? 13.4A

1 and **2** look at ways of verifying that the exterior angles of a polygon sum to 360°.

3 moves on to using this result with regular polygons. Students may need reminding about the meaning of the word "regular".

Practice 13.4B

1 is a useful discussion of how to form the triangles inside a polygon in order to derive the result for the angle sum. Help students to see that drawing triangles from one vertex only creates angles in triangles that will have the same total as the polygon, and no extras. Question 2 also illustrates this.

3 to **6** continue the development of the standard result for the angle sum of a polygon.

What do you think? 13.4B

These questions encourage students to derive the formula for the angle sum of an n-sided polygon, which is then quoted in Example 4.

Practice 13.4C

1 checks that students can substitute correctly into the formula.

2 and **3** develop students' understanding by using the angle sum of a polygon to find unknown angles.

4 to **6** work with regular polygons and both of the rules discovered in this chapter.

7 works backwards from a given angle sum. You may wish to give more totals and challenge students to find the number of sides of each polygon.

What do you think? 13.4C

1 provides more practice at working backwards from a given exterior angle. Some students may not use this for part b, so you could compare finding the number of sides from the formula for the sum of the interior angles with using the formula for the exterior angles.

2 Students may need guiding to use the word "factor".

Consolidate – do you need more?

This revisits the rules and provides some fluency practice.

Stretch – can you deepen your learning?

1 You may wish to consider alternative forms, and show that they are equivalent.

2 Again, factors of 360 are important. There are many online demonstrations that all quadrilaterals will tessellate the plane.

Reflect

The reflection checks understanding of the key vocabulary and learning of the chapter.

13.5 Geometric proof

Small steps

- Solve complex problems with angles in parallel lines
- Prove simple geometric facts (H)

Rationale

Before looking at the demanding Higher step of formal proof, this chapter checks students' familiarity with the required rules by looking at increasingly complex missing-angle problems. Encourage students to write on copies of diagrams while working through their solutions, and to record step-by-step what they have found, giving reasons at each stage. This process is required for the formal proofs that follow. The exercises concentrate on the principle rather than complexity of proof, and many results need only a few lines of reasoning. Replacing letters with numbers in the proof questions and first working through a numerical version is a good starting point for developing the abstract proof versions of the answers. The learning in this chapter is built on in Block 9 of Book 3.

Things to look out for

- Students mixing up the rules they have learned – encourage students to work slowly and systematically, and emphasise the need for accuracy and reflection rather than speed
- As the problems are complex, solutions are not always obvious and students may need support to get started – ask questions like "What can we find out first?" to help students work their way to a solution. Students could also start by listing (mentally or otherwise) the possible rules that could be used in a given diagram, and then they can decide which will be useful for the problem in hand.
- Difficulty simplifying expressions with negative terms, e.g. in writing $180° - x - y$ as $180 - (x + y)$ – link back to expanding brackets and directed numbers

Interleaving opportunities

- All the rules from earlier in this block are used and considered more formally
- The three-letter naming convention for angles is needed for some of the problems and for the formal proof
- Use of algebraic expressions is included, e.g. $180° - (x + y)$
- Skills associated with directed numbers, equivalent expressions and brackets will help when working with expressions involving negative coefficients

Models and representations

- All the models used in previous chapters apply here. In particular, dynamic geometry software is useful but, once again, students may need reminding that a demonstration is different from a proof.

Behind the questions

Are you ready?

These questions assess whether students are confident with the angle rules they have already studied. Any issues here mean that students will require further practice/intervention before moving on to this more demanding chapter.

Practice 13.5A

1 provides an introduction to step-by-step deduction for finding missing angles. You may wish to use a visualiser to compare students' solutions. Discuss how to set out a solution clearly, and explain how this can help to find the answer.

2 Students will need to add letters to explain their working clearly. Even if students are finding this difficult, they should be able to deduce the numerical values of many angles, which is a good starting point.

3 reminds students that the rules for angles in parallel lines apply to sets of parallel lines, not just one pair.

4 to **6** gradually increase the complexity.

What do you think? 13.5A

1 This question becomes much easier once the right angle is added to the diagram.

2 Again, a step-by-step approach should bear fruit.

Practice 13.5B

1 The idea of proof is introduced through a numerical example.

2 Likewise, a particular case is introduced before looking at a general case. Students may need help to see the link between the two parts, and with the negative terms.

3 uses the rules for angles on parallel lines to prove that the angle sum of a triangle is 180°.

4 uses the technique from question 2, and looks a specific case before the general result,

5 You could suggest to students that they pick some numerical values of *a* and *x* before attempting the proofs.

What do you think? 13.5B

These more complex questions would also benefit from trying numerical values first.

Consolidate – do you need more?

This section focuses on the problem-solving rather than proof. There are more proof questions in Chapter 12.6 of Book 1 if required.

Stretch – can you deepen your learning?

1 links to similar shapes as studied in Block 2.

2 and **3** require competence at solving equations as well as using angle facts.

Reflect

The reflection questions are good discussion points. You may wish to hold the discussion first and then ask students to record a summary of this.

13.6 Constructing bisectors (H)

Small steps

- Construct an angle bisector (H)
- Construct a perpendicular bisector of a line segment (H)

Rationale

Students need to be skilled at the accurate use of a pair of compasses to perform these constructions well. If there are difficulties, you may wish to invest more time with the basic constructions of Chapter 13.3 and leave this Higher step until Book 3, where the methods are again covered in detail, alongside the other standard constructions.

Things to look out for	Interleaving opportunities
• Mixing up angle and line bisectors – discuss the stages of the constructions and how they relate to points equidistant from a given point, etc. • Inaccuracy	• This builds on the study in Chapter 13.3 • The perpendicular bisector is linked to the diagonals of a rhombus • The techniques could be revisited when looking at symmetry in Block 15

Models and representations

- You can make sense of the loci of the constructions and the idea of equal distances by using coins, counters, etc.

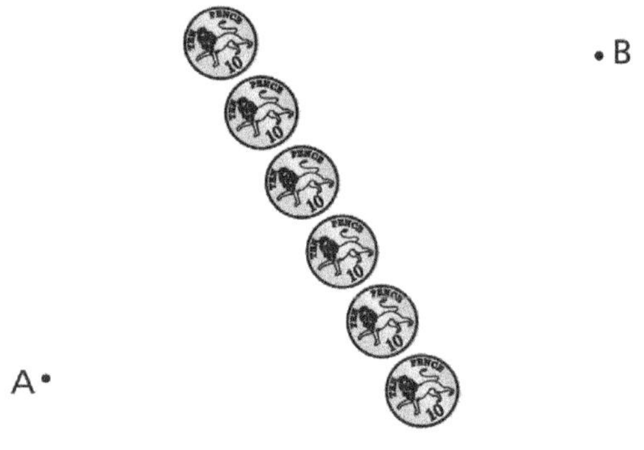

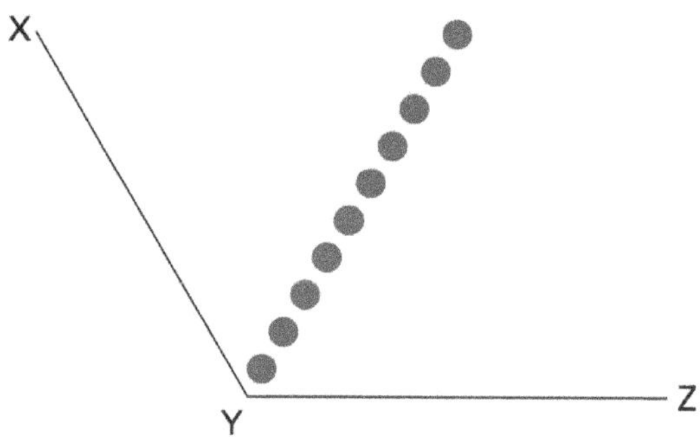

- You can also use "human geometry" by asking students to stand at an equal distance from, for example, two walls in the school hall.

Behind the questions

Are you ready?

Here students practise basic skills of using a pair of compasses.

Practice 13.6A

1 and **2** provide practice at using the technique demonstrated in Example 1. You could model this yourself with a visualiser or use online videos.

3 explores the meaning of "equidistant".

4 provides practice in a context.

5 challenges a possible misconception.

What do you think? 13.6A

To construct an incircle accurately, you need a large triangle, otherwise it becomes very difficult. Looking at a variety of triangles, as in question 2, gives students plenty of practice.

Practice 13.6B

1 and **2** are again practice at using the technique demonstrated in Example 2. Again, you could model this yourself with a visualiser or use online videos.

3 uses both angle and perpendicular bisectors, so students have to reflect on when to use each technique.

4 illustrates why the technique works. Students are asked to construct a rhombus (without drawing its sides) using the fact that the diagonals of a rhombus bisect each other at right angles.

What do you think? 13.6B

A circumcircle is slightly easier to draw than an incircle, but care needs to be taken to ensure that it fits on the page as its extent is not entirely predictable.

Consolidate – do you need more?

This extra practice could be carried out at any time to help students to remember the learning.

Stretch – can you deepen your learning?

1 reinforces the understanding of “equidistant”.

2 looks at the properties of diagonals.

Reflect

Students can record the learning in their own words.

14.1 Area of triangles and quadrilaterals

Small steps

- Calculate the area of triangles, rectangles and parallelograms (R)
- Calculate the area of a trapezium

Rationale

In this chapter, students calculate the areas of triangles and quadrilaterals. All students should be familiar with calculating the areas of triangles, rectangles and parallelograms from Book 1. This step allows them to recap this knowledge and look at it in more detail so that they can apply it to finding areas of compound shapes later in the block. Calculating the area of a trapezium was a Higher step in Book 1 so some students will have seen this before whilst to others it may be new. Students should be able to identify the parallel sides of a trapezium and to calculate the area either by substituting into the formula or by splitting it into simpler shapes.

Things to look out for	Interleaving opportunities
• Not dividing by 2 when calculating the area of a triangle • Using the slant height rather than the perpendicular height of a shape • Adding the wrong sides when calculating the area of a trapezium • Not noticing where different units are used within the same question	• Names of shapes • Properties of shapes • Substitution • Algebraic notation • Solving equations

Models and representations

- A geoboard can be used to support understanding of area. Students could make shapes and explore how the dimensions affect the area.

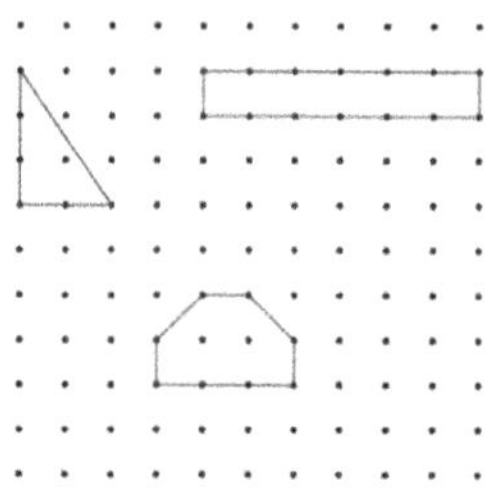

Behind the questions

Are you ready?

These three questions cover the essential prerequisite knowledge for this chapter. Question 2 reminds students that area is the space inside a 2-D shape, which can be seen by counting squares. Question 3 introduces students to the type of calculation they will need to perform later in the chapter. They should recognise that parts b and c are equivalent and should be able to articulate why this is.

Practice 14.1A

1 provides practice at calculating areas. Notice that only the relevant side lengths are given in each diagram. This means that students do not need to think about which lengths to choose at this stage; instead, they can focus on which shape they are dealing with and the method needed to calculate its area. Students could be encouraged to articulate how they know that the lengths that are given are the correct ones to use. Watch out for part f, where students need to recognise that the given shape is a square.

2 requires students to reflect on the difference between calculating the areas of a rectangle and a triangle.

3 Here students need to think carefully about which side lengths to choose when calculating the area. In part b, they are prompted to recognise that some students do not realise this.

4 and **5** interleave prior learning on decimals, fractions and algebra. Although this might look more challenging, students should realise that the method for calculating the area remains unchanged. In question 5, students should discuss their methods in detail. For example, for part a i some students might calculate 4×4.5 and then divide by 2, but this means calculating with decimals twice. Instead, they could do 2×4.5, to make it easier.

7 looks at calculating unknown sides by forming and solving an equation. This method will get students into good habits by encouraging them to think about what they know, and how they can use this to find out information about the shapes.

8 Some students may just divide the area by the side length and not use the formula. If any students do this, they could be encouraged to look back at question 7 and use Samira's method.

What do you think? 14.1A

These two questions are designed to make students think more deeply about how to find the area of quadrilaterals. In question 2, once students have calculated the area of one small rectangle, a bar model could help them to work out its length and width.

Practice 14.1B

1 Here the focus is on identifying the parallel sides of each trapezium. This is an essential first step before calculating the area. Students should notice that although the edges aren't marked with "parallel" arrows, they can identify them from the right angles shown in the diagrams.

2 Students use the trapezia from question 1 but with side lengths given. They have already identified the parallel sides, so they can now focus their attention on substituting the correct values into the formula for the area.

3 Students could sketch a diagram to help them to answer this question.

4 Watch out for students using the incorrect height in part a. In part b, students should notice that the lengths are not all given in the same units, so they will need to convert before they can calculate the area.

5 This is similar to Samira's method in Practice 14.1A. Students could be encouraged to write down the formula, substitute in the values they know and then solve an equation to find the missing lengths.

What do you think? 14.1B

These two questions provide the opportunity for students to think about the area of a trapezium in more detail. Question 2 can be used to link the formula for the area of a trapezium to those for a rectangle and parallelogram, whilst exposing why a more general version of the formula is necessary.

Consolidate – do you need more?

These questions revisit the key points covered in the previous exercises, to ensure that students can calculate the areas of triangles and quadrilaterals.

Stretch – can you deepen your learning?

1 highlights the fact that it doesn't matter which side of a shape you take to be the base, as long as the height is perpendicular to it.

2 Notice how the numbers used in each part are the same. Students should focus on how the position of each number affects how they need to tackle the question.

3 Students should reflect on how they can see that b is longer than a by looking at the diagram. To increase the level of difficulty, they could consider how the values of a and b would be affected if they didn't have this information.

4 Here students should draw a diagram to support their thinking. Part a looks back at learning from Block 7 on expanding binomials. Students haven't yet solved a quadratic equation of this form, so part b expresses it in such a way that they can solve it. Students should be prompted to consider why -9 is not a possible value of x.

Reflect

Students should explain their learning in their own words as this will clarify their thinking. They should be able to explain clearly how to calculate the areas of triangles and parallelograms, and sketch shapes that have a given area.

14.2 Area of a circle

Small steps

- Investigate the area of a circle
- Calculate the area of a circle and parts of a circle without a calculator
- Calculate the area of a circle and parts of a circle with a calculator

Rationale

In this chapter, students learn how to calculate the area of a circle. Practical activities could be used to investigate this and to support understanding of the formula prior to working on this chapter. Guidance on these sorts of activities can be found in the White Rose Maths schemes of learning. This chapter gives students plenty of opportunity to calculate areas of circles where the questions are presented in a variety of ways. The first part of the chapter looks at questions where answers are given in terms of π. This means that students don't need to think about rounding during this section and can focus fully on the new skill they are learning. In the second part, they are asked to give their answers to a specified number of decimal places. It is important that students understand that leaving answers in terms of π is more accurate as there is no rounding.

Things to look out for	Interleaving opportunities
• Using the radius instead of the diameter, or vice versa • Not squaring the radius • Multiplying by 2 rather than squaring • Dividing the radius by 2 when finding the area of a semicircle rather than dividing the area by 2 • Rounding errors	• Areas of triangles and quadrilaterals • Algebraic notation • Substitution • Rounding • Square numbers • Solving equations • Calculator skills

Models and representations

- Linking the area of a circle to the area of a parallelogram can help students to understand the formula for the area of a circle. The more sectors the circle is divided into, the closer to a parallelogram the shape becomes.

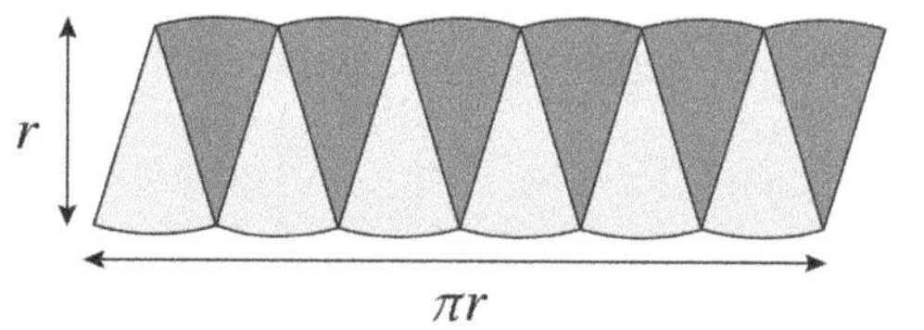

Behind the questions

Are you ready?

These six questions cover the prerequisite skills needed to complete this chapter. Question 1 ensures that students are familiar with the names of the parts of circles, whilst questions 2 to 5 formalise this by focusing specifically on the radius and the diameter. It is important that students not only know that the radius is half of the diameter, but also that these lines must either go through, or stem from, the centre of a circle. Question 6 gives students the opportunity to recap rounding, which will prepare them for the questions in Practice 14.2B.

Practice 14.2A

1 In all the diagrams, the length given is the radius, so students just need to square this and give their answer in terms of π. Notice how the value of the radius in part d is one tenth of the radius value in part c (ignoring the units). Students may be able to square 13 but might struggle to square 1.3. However, if this link is made explicit it can help them with the calculation.

2 highlights the common error of calculating the area using the diameter rather than the radius. It is important that students are exposed to this early in the chapter so that they don't make this mistake.

3 Students are now exposed to examples where they have to calculate the area given the diameter. If any students square the diameter, they should be referred back to question 2. Watch out for students halving the square of the diameter rather than squaring half of the diameter.

4 Students now look at parts of circles. They should notice that each of these images shows the same circle, but a different amount of it. They should be able to explain how they know that part c is a quarter of a circle, i.e. because 90° is one quarter of 360°.

6 Students are required to find the radius given the area. Notice how all the numbers used are square numbers, so as to simplify the calculations. Students could be encouraged to write out the formula and substitute the value they know so as to form and then solve an equation.

What do you think? 14.2A

These three questions are designed to make students think about the areas of circles in more detail. In question 1, they should recognise that the diameter of the circle must be equal to the width of the rectangle. In question 2, watch out for students squaring only 7 or y and not squaring the whole term.

Practice 14.2B

1 allows students to practise finding the area of a circle using a calculator. Some parts give the radius and some the diameter. Students should now be able to select the correct length to use in the calculation.

2 Notice how part b is the same as part b of question 5 in Practice 14.2A. Students should recognise that the decimal answer and the answer in terms of π are equivalent.

4 Students could be prompted to write out the formula and substitute the value they know.

5 When working out the radius, students will find that it is 2.999 cm. They should be able to explain why drawing a circle of radius 3 cm is fine for this.

6 Watch out for students calculating the radius of the pond and not multiplying this by 2 to find the diameter.

What do you think? 14.2B

These four questions allow students to calculate the area of a circle in different contexts. In question 1, students might be tempted to think that two 9-inch pizzas have the same area as one 18-inch pizza and should be encouraged to investigate this by considering how many times bigger the 18-inch pizza is than a 9-inch pizza. This could be explored for other sizes too, so that students realise that if the diameter of one pizza is double another, then its area will be four times as large.

Consolidate – do you need more?

These questions revisit the key points covered in the previous exercises, to ensure that students can calculate the area of a circle both with and without a calculator.

Stretch – can you deepen your learning?

1 Watch out for students giving the cost for a decimal value for the number of tins instead of realising that this must be an integer.

2 Students could be encouraged to form and solve an equation to answer this question.

3 interleaves prior learning on ratio and the area and perimeter of rectangles. Some students may work out the area of the rectangle to be 100 cm^2 and then share this in the ratio 4 : 1 rather than considering what the side lengths should be.

Reflect

Students should explain their learning in their own words as this will clarify their thinking. They should be able to articulate why an answer given in terms of π is more accurate than a decimal answer, and also recognise and explain common misconceptions.

14.3 Area of compound shapes

Small steps

- Calculate the perimeter and area of compound shapes

Rationale

Here, students bring together their learning from the previous chapters to calculate the areas of compound shapes. They also revises perimeter, as covered in Book 1. Students should be aware that a compound shape, also known as a composite shape, is a shape that is made up of two or more other shapes. So although at a glance students may think that they don't know how to calculate the areas of these shapes, they should be encouraged to break them down into smaller shapes that they are familiar with and whose areas they know how to find. The first prt of the chapter focuses on rectilinear shapes, which will require students to find missing lengths before they can calculate the area; some students may require assistance with this. The second part looks at composite shapes made up of triangles and circles. Mental and written methods of calculation should be encouraged where possible, but calculators could be used throughout, if necessary, to ensure that the calculations don't form a barrier to the new learning.

Things to look out for	Interleaving opportunities
• Multiplying all lengths given • Not calculating missing lengths • Incorrectly calculating missing lengths • Subtracting/adding the areas when they should be added/subtracted • Using a slant height rather than a perpendicular height • Trying to work out too much in one step • Thinking that the total perimeter of a compound shape is the sum of the perimeters of its constituent parts	• Addition and subtraction • Multiplication and division • Areas of rectangles, triangles and parallelograms • Area of a trapezium • Area of a circle • Percentages • Algebraic notation • Expanding double brackets

Models and representations

- Colour can be used to support understanding of finding missing side lengths. It helps students to see which sides are linked and the calculation that needs to be carried out.

The two shorter vertical blue side lengths add together to make the longer vertical blue side.

The two shorter horizontal yellow side lengths add together to make the longer horizontal yellow side.

Behind the questions

Are you ready?

These three questions cover the essential knowledge for this chapter. Questions 1 and 2 revisit calculating the areas of triangles, quadrilaterals and circles, which were covered earlier in this block. It is important that students can confidently calculate the areas of these shapes before moving on to the content of this chapter. Although it may not be immediately obvious why question 3 is linked to the lesson, it will help students to find missing side lengths. As well as being able to work out the missing value in each diagram, it is important that they can articulate why they chose to carry out each calculation. For example, in part a, they know that the two parts form the whole, so they need to total 3 and 21 to give 24. This could then be applied in the context of shape if the two shorter sides were 3 and 21, and the longer parallel side was 24.

Practice 14.3A

1 Here the focus is on finding missing lengths. The use of colour can help students with this, if necessary. As well as being able to find the unknown lengths, it is important that students can articulate why they chose to perform each calculation.

2 Notice how the shapes are identical to those given in question 1. Students should use the side lengths they have just found in their calculations for each area.

3 Students are encouraged to think about the different ways of calculating the area of a compound shape. Although it isn't always the case that the quickest way is best, it is important to encourage students to use the most efficient method and to be able to think flexibly.

4 Students should reflect on the difference between this question and the previous ones, to realise that, rather than finding the total of their areas, here they need to use subtraction. Students could think of the shapes being "cut out" to support their understanding of why subtraction is necessary.

What do you think? 14.3A

Question 1 looks at area of compound shapes in a more challenging context, revisiting content from earlier in the curriculum. Students should be encouraged to write down each step of their method to ensure that they don't make mistakes. They could draw the rectangles separately and focus on finding missing side lengths first.

Practice 14.3B

1 Here each shape can be split into a rectangle and triangle(s). Although the dimensions of the rectangle are quite obvious, students should be encouraged to explain how they know the base and perpendicular height of the triangles. Watch out for students not dividing by 2 when calculating the area of a triangle.

2 looks at splitting a compound shape in different ways. Although students are encouraged to choose which method they prefer, they should also appreciate that each is valid and be able to use both.

3 provides practice at finding the areas of compound shapes. Students should be encouraged to sketch each shape and label each dimension clearly to support their working.

4 can feel quite overwhelming at first glance. Students could be prompted to consider whether they actually need to work out the area of the patio, or whether they can find the answer without it. Some students could be confused by the context if they are not familiar with words such as "patio", but they should try not to let this distract them from the maths.

What do you think? 14.3B

These questions look at areas of compound shapes in different ways. In question 2, students should recognise that the formula for the area of a triangle states that it is half of the area of the rectangle in which it is enclosed, so no calculation is needed to know that the answer is 50%.

Consolidate – do you need more?

These questions revisit the key points covered so far, to ensure that students are able to calculate the areas of compound shapes. In question 3, watch out for students giving the area of the bigger circle as the answer for part b rather than subtracting the area of the smaller circle.

Stretch – can you deepen your learning?

1 Students need to recognise that the unshaded triangle is isosceles, which will enable them to work out the base of the shaded triangle.

2 interleaves prior learning on algebra and expanding a pair of binomials. Students could use a bar model to help them to find missing side lengths when dimensions are given as expressions.

3 When working out the perimeter, ensure that students do not include the internal horizontal width at the top of the rectangle. Part c is an investigation into how shapes with the same area can have different perimeters.

Reflect

Students should explain their learning in their own words to help to clarify their thinking. They should be able to split compound shapes in several ways and be able to choose the most appropriate method for each question.

15.1 Line symmetry of shapes

Small steps

- Recognise line symmetry

Rationale

Students should already be familiar with line symmetry from earlier in the national curriculum, but this chapter can be used to revisit this and also to extend it to consider it more generally. Students are likely to be able to recognise line symmetry in shapes but they might not realise that there is a connection between the type of shape and its key features and the number of lines of symmetry it has. For example, they might not realise that the number of lines of symmetry of a regular polygon is equal to the number of sides. In this chapter, they can explore this and make generalisations and predictions.

Things to look out for

- Drawing a straight line through the middle of a shape even though it is not a line of symmetry
- Assuming that the number of lines of symmetry is equal to the number of sides in *all* shapes
- Thinking that the diagonals of a rectangle are lines of symmetry
- Not considering how the pattern on a shape affects the number of lines of symmetry

Interleaving opportunities

- Names of shapes
- Properties of shapes
- Angles in polygons
- Types of triangles
- Types of quadrilaterals
- Two-way tables

Models and representations

- Mirrors and paper folding are useful tools when modelling line symmetry. They help students to understand what is meant by a line of symmetry and are great way of testing whether or not a line on a shape is actually a line of symmetry.

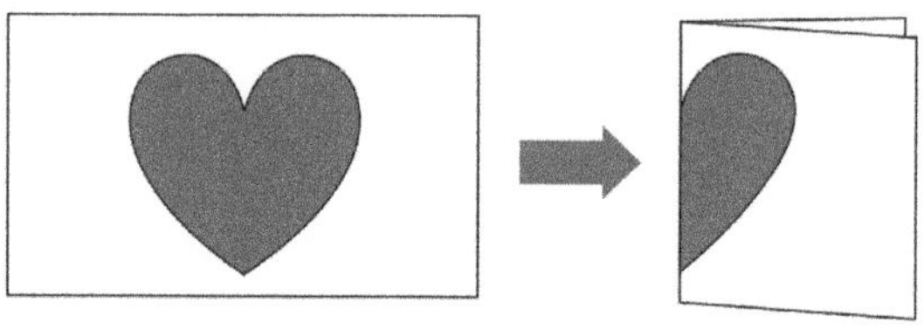

Behind the questions

Are you ready?

These two questions provide opportunities for students to practise the key skills needed in this lesson. Question 1 ensures that students are familiar with the mathematical names of shapes, and also with their properties. In part c, some students might say that only polygon iii is isosceles because they're not familiar with an isosceles trapezium. They could think about this as an isosceles triangle "with the top cut off" to support their understanding of why it is isosceles. Question 2 ensures that students are familiar with the word "symmetrical" and what it means.

Practice 15.1A

1 provides students with the opportunity to identify lines of symmetry in common 2-D shapes. Students might think that the parallelogram has lines of symmetry, and they could be encouraged to use a mirror to aid their understanding of why it doesn't. Students could be stretched to consider why it is important that the triangle is isosceles.

2 Here students have to identify which lines of symmetry are drawn correctly. As well as being able to identify this, it is important that they can verbalise their reasoning and explain why they think that each line is or isn't a line of symmetry.

3 gets students thinking about how the number of lines of symmetry of a regular polygon is linked to the number of sides. They should notice that the number of lines of symmetry is equal to the number of sides and consider why this happens.

4 quickly addresses the misconception that arises when students generalise too quickly. They have just seen the connection between the number of sides and the number of lines of symmetry of *regular* polygons and it's important that they understand that this only applies to regular polygons. This question quickly highlights and unpicks that misconception.

5 Students could be encouraged to make predictions about the number of lines of symmetry before drawing the shapes to test them. Some students might be tempted to say that a semicircle has an infinite number of lines of symmetry because a circle does, but drawing the shape will help them to understand why this isn't the case.

6 encourages students to relate lines of symmetry to the angles in a shape. Students should realise that in order for a shape to have a line of symmetry, there must be at least one pair of equal angles. As this triangle is isosceles, it must have one line of symmetry. This question could be used to revisit constructions from Book 1, and students could accurately draw a triangle with these angles to check their answers.

What do you think? 15.1A

These three questions encourage students to think about lines of symmetry more deeply. Question 1 links to the misconception explored in question 4 of Practice 15.1A, but without any shapes on which to base their answers. Questions 2 and 3 look at how patterns on shapes affect the number of lines of symmetry. In question 3, students could be extended to consider which letter could be placed on logo 2 so that the number of lines of symmetry remains unchanged.

Consolidate – do you need more?

These questions revisit the key points covered in the previous exercise and ensure that students can not only recognise lines of symmetry in a shape, but can also recognise when a line of symmetry is drawn incorrectly. Question 3 revisits two-way tables and gets students to sort shapes according to their properties.

Stretch – can you deepen your learning?

1 Students should be careful in part c that they reflect each part of the pattern in *both* lines of symmetry: watch out for students not fully completing this part. In the discussion task at the end, students should notice that although it is possible for a square to have 1, 2 or 4 lines of symmetry, it is not possible for it to have 3 lines of symmetry. This means that there isn't a pattern that can be drawn on a squared grid that gives 3 lines of symmetry.

3 generalises the idea that a regular polygon has the same number of lines of symmetry as its number of sides.

4 Here students use the fact that because there isn't a pair of equal angles, there can't be a line of symmetry. Some students might think that there is one because the angles are close in size, but they should realise that because they are not equal, there isn't a line of symmetry.

Reflect

Students should explain their learning in their own words as this will clarify their thinking. They should be able to explain clearly what is meant by a line of symmetry and give examples of shapes that do and do not have them.

15.2 Reflecting in horizontal or vertical lines

Small steps

- Reflect a shape in a horizontal or vertical line

Rationale

In this chapter, students build on their knowledge of lines of symmetry to reflect shapes in horizontal and vertical lines. They start by focusing on examples where the shape is touching the line. This can be compared with completing a symmetrical figure. Once students are confident with this, they will look at examples where the shape is not touching the line. Here, they not only need to consider what happens to the shape but also the distance between the shape and the mirror line. This step provides the opportunity to revisit writing equations of lines parallel to the axes, which was covered earlier in the year. Students will be expected to reflect shapes in lines of the form $y = a$ and $x = b$.

Things to look out for	Interleaving opportunities
• Drawing the image on the other side of the line of reflection without actually reflecting the object • Drawing the image so that an edge is on the line of reflection when it shouldn't be • Incorrect plotting of the lines $y = a$ and $x = b$ resulting in an incorrect reflection • Drawing an incorrect reflection when a shape crosses one of the axes • Incorrect drawing of the shape	• Properties of shape • Coordinates • Equations of lines parallel to the axes • Area

Models and representations

- Student need to understand the language used in this chapter.
  The original shape is called the object, and the reflected shape is called the image.

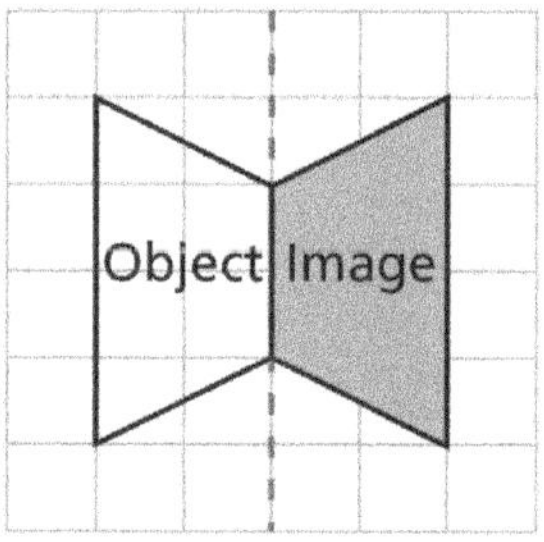

Behind the questions

Are you ready?

These three questions cover the essential knowledge for this chapter. Question 1 revisits content from the previous chapter on line symmetry. Questions 2 and 3 recap learning from earlier in the year on working in the Cartesian plane. Question 2 checks that students are able to read coordinates whilst question 3 looks at lines parallel to the axes.

Practice 15.2A

1 Here the shapes are touching the mirror line and the reflection isn't on a coordinate grid. This can be linked to the idea of completing a symmetrical figure. Students could use a mirror to help them with this question.

2 Following on from the previous question, students now look at shapes that aren't touching the mirror line. Again, a mirror could be used, and students should be encouraged to describe the differences between the positions of the object and the image in these questions and those in question 1.

3 looks at common misconceptions within this topic. Students should be able to identify a correct reflection as well as explain the mistake that has been made in any incorrect ones. Identifying and exposing these misconceptions early on can help to prevent students making the same mistakes later.

4 Notice that a coordinate grid has been subtly introduced here so that students can appreciate that reflecting on a coordinate grid involves exactly the same process as reflecting on plain squared paper. As well as reflecting shapes on a coordinate grid, students are also required to write the coordinates of each image. As an extension, they could be asked to consider the relationship between the equation of the line of reflection and the coordinates of the corresponding object and image.

5 Although students are not necessarily required to copy and reflect these shapes, you could suggest that they do. As with question 4, students could be extended by considering the relationship between the line of reflection and the coordinates of the corresponding object and image.

6 and **7** Students could be encouraged to draw diagrams and use tracing paper. As well as being able to draw or identify the position of the line of reflection, they also need to give its equation. Writing down some coordinates on the line of reflection would help them to deduce this.

8 Often, when a reflection crosses an axis, students miscount the squares and draw the image incorrectly. This question is designed to expose this common mistake by asking students to explain the error.

What do you think? 15.2A

These two questions require students to think more deeply about reflecting shapes. In question 1, students could be encouraged to make predictions before drawing a diagram to work out their answers. They could also be asked, "What if Chloe was reflecting in the *y*-axis instead?" When making predictions, students should consider what will stay the same and what will change from the object to the image. In question 2, the object and image overlap.

Consolidate – do you need more?

These questions revisit the key points covered in the previous exercise, to ensure that students can reflect a given shape, as well as identify mistakes that have been made in any incorrect reflections.

Stretch – can you deepen your learning?

1 introduces the idea of invariant points, without actually using that vocabulary. Students could highlight each vertex of the triangle in a different colour to help them to understand why all three points are not invariant.

2 interleaves learning from Chapter 14.1 on the area of a trapezium. Some students may be confused about the lack of units; as there aren't any standard units of measure involved, students can just give their answers in terms of "square units".

3 As an extension, students could be asked to consider a reflection that would result in different lines bordering the rectangle.

Reflect

Students should explain their learning in their own words as this will clarify their thinking. They should be able to explain clearly what is meant by a line of symmetry and give examples of shapes that do and do not have them.

15.3 Reflecting in diagonal lines

Small steps

- Reflect a shape in a diagonal line

Rationale

Here, students build on their learning from the previous chapters by reflecting shapes in diagonal lines. This can cause confusion as some shapes appear to change orientation when reflected in a diagonal line. The use of tracing paper can support understanding. Also, you could suggest that students rotate their pages so that the mirror line becomes horizontal or vertical to make it easier to work out and check the answers.

Things to look out for	Interleaving opportunities
• Treating a diagonal line as a horizontal or vertical line and just "completing" the shape • Counting only horizontally or vertically to a diagonal line and hence drawing an incorrect reflection • Not up lining tracing paper correctly after turning it over • Incorrect use of the lines $y = x$ and $y = -x$	• Properties of shape • Coordinates • Equations of lines of the form $y = kx$

Models and representations

- Students need to understand the language used in this chapter. The original shape is called the object, and the reflected shape is called the image.
- When reflecting in diagonal lines, tracing paper can be particularly useful. Ensure students have tried both using tracing paper and counting squares diagonally so that they can decide which method works best for them.

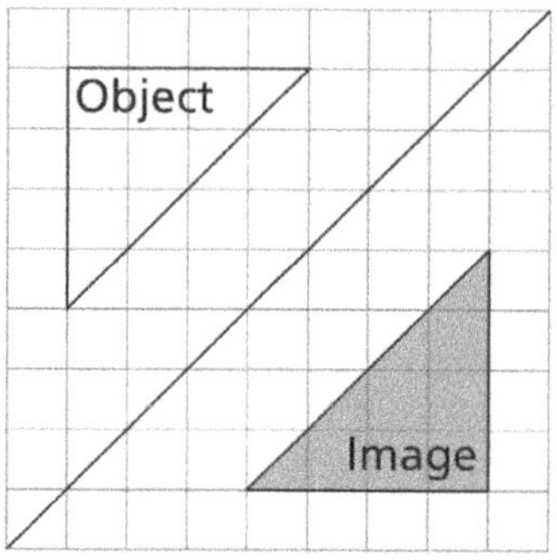

Behind the questions

Are you ready?

These three questions cover the essential knowledge for this chapter. Question 1 looks at lines of symmetry and it is crucial that students fully understand why diagram B does not show a correct line of symmetry. This will prepare them for what follows in the rest of the chapter. Questions 2 and 3 remind students of the line $y = x$ as this will be used later in the lesson as a mirror line.

Practice 15.3A

1 Here, shapes are touching the mirror line and the reflection isn't on a coordinate grid. Parts a and b can be linked to the idea of completing a symmetrical figure. Students may wish to use a mirror for support.

2 Following on from the previous question, students now look at shapes that aren't touching the mirror line. Notice that the shapes are the same as in question 1; the only difference is that the shapes aren't touching the line of reflection. Before completing these questions, students could be encouraged to make predictions about what they think will be the same and what will be different about their answers to questions 1 and 2. They should recognise that although the orientation of the image will be the same, its position relative to the line of reflection will be different.

3 and **4** look at common misconceptions when reflecting in diagonal lines. As well as being able to identify that a reflection is incorrect, students should be able to explain the mistake that has been made. This will help to prevent them making the same mistakes themselves.

5 Students are required to plot the shape before reflecting it. As an extension, they could consider the relationship between the equation of the line of reflection and the corresponding coordinates of the object and the image.

What do you think? 15.3A

These two questions require students to think more deeply about reflecting shapes. In question 1, the shape is overlapping the diagonal line of reflection. Students could complete this reflection themselves to support their reasoning when answering the question. Before completing the reflection in question 2, students could make predictions about the coordinates of the image.

Consolidate – do you need more?

These questions revisit the key points covered in the previous exercise, to ensure that students can complete a reflection in a diagonal line, both when the shape is touching the line of reflection and when it isn't.

Stretch – can you deepen your learning?

1 requires students to identify the line of reflection, which revisits content from earlier in the year. Notice that horizontal and vertical lines are included to provide students with practice at distinguishing between the two. Students could be encouraged to draw each line of reflection before working out its equation. Students could write down the coordinates of some points on each line to help them to deduce the equation.

2 looks at reflecting lines. As an extension, students could consider other similar situations: for example, can they generalise the result if the line $y = a$ is reflected in the line $y = x$? What if the line of reflection was $y = 2x$ or $y = x + 2$?

Reflect

Students should explain their learning in their own words as this will clarify their thinking. They should be able to explain clearly how to reflect a shape in a diagonal line, and should know the similarities and differences between the reflections in this chapter and those in Chapter 15.2.

16.1 Collecting data

Small steps

- Set up a statistical enquiry
- Design and criticise questionnaires

Rationale

The purpose of this chapter is to look in depth at collecting data. The first part of the chapter focuses on setting up a statistical enquiry. Students will need to create a hypothesis to test using both primary and secondary data. They should understand that a hypothesis is an idea that they want to investigate further and is always posed as a statement and not as a question. In the second part of the chapter, students will design and criticise questions on questionnaires. It's important that they can identify things that are wrong with questions and are able to improve them. Students should understand what is meant by a leading question, and know how to pose a question in a way that doesn't suggest their own personal opinion.

Things to look out for

- Not covering all possible responses with the answer options
- Using a question as a hypothesis
- Overlapping the categories in the answer options to a question
- Not giving a time frame in a question
- Writing leading questions

Interleaving opportunities

- Types of data
- Tally charts
- Inequality symbols

Models and representations

- Tally charts are used to collect and record data.

Method of travel	Tally	Frequency
walk	~~||||~~ ||||	9
bike	|||	3
bus	~~||||~~ ~~||||~~ ||	12
car	~~||||~~		6

Behind the questions

Are you ready?

These two questions allow students to practise the key skills needed in this chapter. Question 1 ensures that students can interpret tally marks and reminds them of the meaning of the word "frequency". It also allows students to recap basic interpretation of data from tables. In question 2, students are reminded of the different types of data, and they are asked to sort data items into the correct categories.

Practice 16.1A

1 and **2** Here students need to write and test a hypothesis. It's important for students to understand that they might come to a different conclusion if they used a different sample. Students could be prompted to explore scenarios that would give a biased sample. For example, would it be a good idea to carry out a survey of how students travel to school by asking students on the school bus?

3 provides an opportunity to revisit learning from Block 5 on scatter graphs.

4 Here students don't need to actually test a hypothesis; instead, they need to think of a suitable one and select a suitable sample size. Students should appreciate that the more data they can collect, the more representative is their sample, but they should also understand the limitations of asking too many people. They could consider an appropriate sample size that would be representative of their whole year group.

What do you think? 16.1A

Here students focus specifically on how the careful choice of a hypothesis and sample can make the data reliable. Students should realise that the hypothesis given here influences people's responses and that only asking his own friends will not produce a good, unbiased sample.

Practice 16.1B

1 Here students discuss whether or not they think the question is good. Initially, some students may find this difficult; they could be prompted to recognise the flaws through questions such as "Mr X hates the menu. Which option does he tick?"

2 Students should be able to transfer what they learned from question 1, where the answer options didn't cover all the possible responses. They should also notice that some possible responses have two boxes, and that the question doesn't fully represent what Zach is trying to find out. He needs to include the words "per month" in the question to make the time frame clear.

3 Students should recognise here that Flo is giving her own opinion within the question and that this makes it a leading question.

4 Students are given the chance to design their own questions to find out the given information. They should reflect on all the errors apparent in the earlier questions as they do this. Students could be encouraged to criticise and improve each other's questions if appropriate.

What do you think? 16.1B

Again, the focus is on the importance of choosing a fair, representative sample, and then designing a suitable question to include in a questionnaire.

Consolidate – do you need more?

These questions revisit the key points covered in the previous exercises to ensure that students can set up a statistical enquiry and design and criticise questionnaires. If more practice is needed, students could be encouraged to work in pairs to create similar questions for each other to answer.

Stretch – can you deepen your learning?

1 Students should consider how the choice of questions will affect Darius's conclusions from his survey.

2 and **3** look at collecting data by means of appropriate samples. Question 3 specifically focuses on different methods of random sampling, and students should consider the pros and cons of each.

Reflect

Students should explain their learning in their own words as this will clarify their thinking. They should be able to explain what is meant by a hypothesis and give examples. They should also be able to explain what makes a good questionnaire, and write a question to support this.

16.2 Lines and charts revisited

Small steps
- Draw and interpret pictograms, bar charts and vertical line charts (R)
- Draw and interpret pie charts (R)
- Choose the most appropriate diagram for a given set of data

Rationale

The purpose of this chapter is to revisit different types of charts and to identify which chart is most suitable for displaying each type of data. Students will have first studied pictograms in KS1 and they will have seen bar charts, line charts and pie charts in KS2 and KS3. Although they should be familiar with all of these types of charts, they may need reminding of the key features, such as leaving a gap between the bars in a bar chart and how to work out the angle of each sector in a pie chart. Students are given plenty of opportunity throughout the chapter to revisit constructing these diagrams. As they do this, they should pay attention to the choice of chart for each type of data. Students should be able to explain why each chart was chosen to display each set of data and whether any other charts would or would not be suitable.

Things to look out for	Interleaving opportunities
• Using frequency as the angle in a pie chart rather than working out the angle • Not putting gaps between bars in a bar chart • Not giving a key with a pictogram • Getting the scale wrong on an axis of a bar or line chart	• Mean • Median • Mode • Four operations • Range • Angles

Models and representations

- Pictograms

- Bar charts

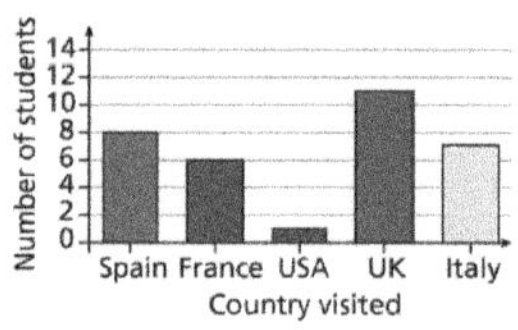

- Line charts

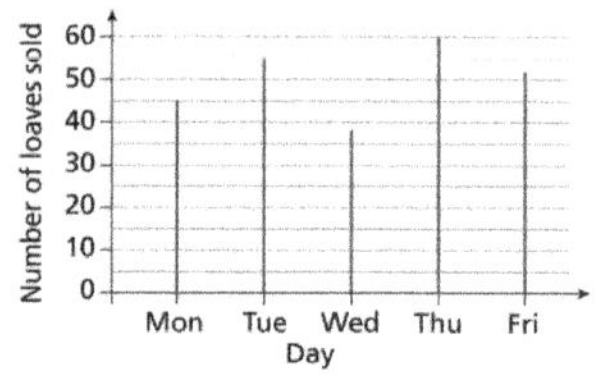

- Pie charts

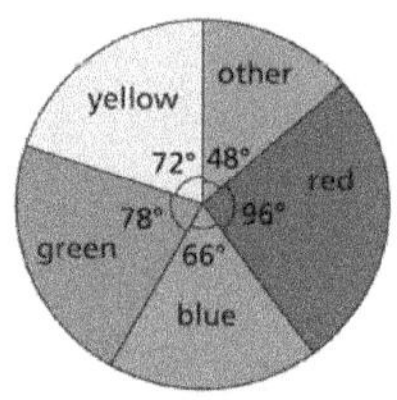

Behind the questions

Are you ready?

These three questions allow students to practise the essential skills needed for this chapter. Question 1 ensures that students can recognise different types of charts, while questions 2 and 3 prepare students for drawing pie charts. Students need to know that the sum of the angles in a full turn is 360°, and they need to be able to draw angles accurately using a protractor.

Practice 16.2A

1 Watch out for students not using the key and just counting the number of “people” in each category. Students should be encouraged to share their methods for calculating the total number of students. For example, some might work out the total for each category whilst others might add up the numbers of “people” and multiply by 4.

2 allows students to criticise a badly drawn pictogram whilst, at the same time, highlighting what makes a good pictogram.

3 Here students can demonstrate what they have learned in question 2 by drawing their own pictograms. Students should explore what would be the same and what would be different if they used different keys. They could explore the pros and cons of different keys to decide which is the most suitable. For example, using a 1 : 1 key would be time consuming, but using a 1 : 100 key would be inaccurate and hard to read.

4 shifts the focus to looking at bar charts. Students should be prompted to notice the different scales used on the axes of the charts. They could be stretched to consider how changing the scale on a chart can make the information appear different from what is actually shown and could be misleading.

5 Students should be encouraged to compare their methods for parts b and c. Some students might choose to work out the total for each destination and then find the difference, or they might find the value from the difference in the bar heights.

6 Students should be encouraged to explain why each "frequency" is being multiplied by 10 to give the angle. They should be aware that that, when drawing the pie chart, their final sector should "complete" the circle and realise that a mistake has been made if this isn't the case.

7 Watch out for students simply using the frequency rather than calculating the angle for each sector. Any students who do this should realise that the circle is not complete and therefore a mistake must have been made. Also watch out for students simply multiplying by 10 because that's what they did in the previous question. They should be prompted to find the total frequency, and then compare this with the total number of degrees. They could think of this as a ratio. The ratio of total frequency to total number of degrees is 40 : 360, so for every 40 people there are 360 degrees, and this can be scaled down to find the angle for each person.

8 This looks at pie charts where the sizes of the sectors are given as percentages rather than angles. Students could be stretched to consider the angle of each sector in the pie chart.

What do you think? 16.2A

These three questions provide opportunities for students to think about charts more deeply. In question 1, students might be tempted to construct a frequency table straight away to represent the counters but, instead, they should try to work out the answer without this step by first eliminating one of the charts and then focusing on the colours in the remaining charts. However, they could use frequency tables to check their answers. Questions 2 and 3 require students to find additional information using what they are given. They should be encouraged to label as much as they can initially, and then fill in any gaps.

Consolidate – do you need more?

These questions revisit the key points covered in the previous exercise to ensure that students can draw and interpret different types of charts. If more practice is needed, students could be encouraged to work in pairs to create similar questions for each other to answer.

Stretch – can you deepen your learning?

1 Watch out for students saying that there is no mode because each age only appears on the chart once. Similarly, they may say that the range is 7 because the highest frequency is 8 and the lowest is 1. It's important that students understand the context in which a question is being asked, and use this to sense-check their answers.

2 asks students to think about the limitations of certain charts. In part b, students should realise that the angles will not work out to be integer values, so these will need to be rounded before a pie chart can be drawn.

3 reinforces the idea that you can't make comparisons about actual numbers by simply looking at pie charts; it is only possible to compare proportions unless the total value is given. Students could be encouraged to construct a frequency chart to support them with part c.

Reflect

Students should explain their learning in their own words as this will clarify their thinking. They should be able to highlight the similarities and differences between different types of charts and know how to draw and interpret each type.

16.3 More lines and charts

Small steps

- Draw and interpret multiple bar charts
- Draw and interpret line graphs
- Represent and interpret grouped quantitative data
- Choose the most appropriate diagram for a given set of data

Rationale

The purpose of this chapter is to build on the previous one by looking at line graphs and more complex types of charts. This chapter has three exercises. The first exercise moves on from the previous chapter to look at multiple bar charts. The second exercise looks at line graphs whilst the third looks at grouped quantitative data. Throughout each exercise, students should be prompted to consider why each type of chart has been chosen to represent the data, and why others wouldn't be suitable for the given context.

Things to look out for

- Drawing bars of different widths
- Using a solid line where it isn't appropriate on a line graph
- Not having gaps between bars in a bar chart
- Using an incorrect scale on an axis
- Not giving a key with a multiple bar chart

Interleaving opportunities

- Averages
- Fractions and percentages
- Two-way tables
- Four operations
- Inequality symbols
- Types of data

Models and representations

- Multiple bar charts

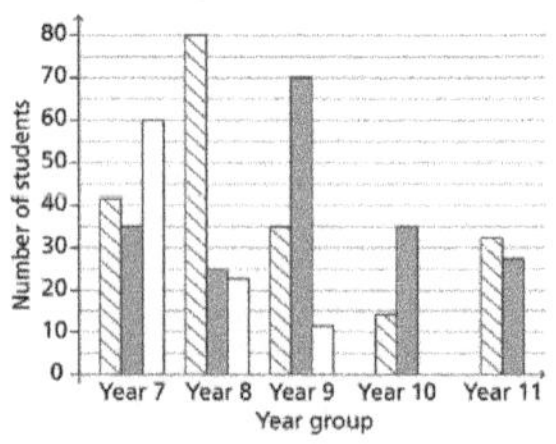

- Line graphs

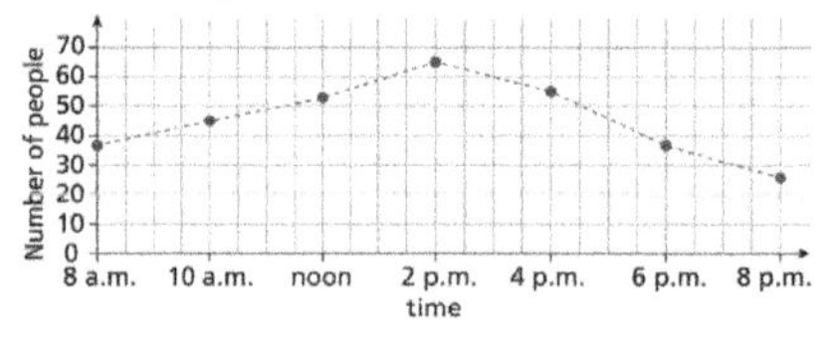

- Frequency diagrams

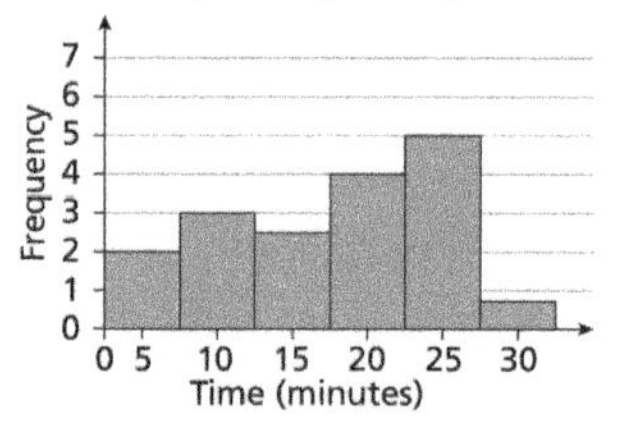

Behind the questions

Are you ready?

Question 1 recaps learning from the previous chapter on interpreting bar charts. Question 2 ensures that students can draw a set of axes accurately and plot points correctly, whilst question 3 checks that students understand the meaning of each inequality symbol.

Practice 16.3A

1 Part a ensures that students can read a scale on an axis correctly. For the later parts, students should be encouraged to compare their methods and decide which is the most efficient.

2 Students could consider how a chart representing the number of points would compare with the given chart.

3 Students could explain why working out is necessary in part a but is also needed for the later parts. You could discuss what can be seen at a glance from a bar chart and what can be found out by deeper analysis.

What do you think? 16.3A

This question interleaves prior learning on two-way tables in the context of multiple bar charts. Students should realise that they could draw this in two different ways, either with the year group on the *x*-axis and a key for the type of meal, or with the type of meal on the *x*-axis and a key to show the year group. Students could be asked to draw the chart in both ways and describe the similarities and differences.

Practice 16.3B

1 Students should understand and be able to explain why their answers to parts c and d are only estimates. They should notice from the graph that the readings were taken only every other day, so any values between these are estimates. Although the line graph shows the trend in change in height, they can't assume that the rate of growth is linear.

2 Watch out for students using a solid line to join the points on the line graph.

3 Students could be stretched to consider what would be different if the data hadn't been recorded on day 3. If this was the case, the graph would give the appearance that it did rain every day. This will help students to understand the limitations of line graphs and the importance of having more data.

4 Students could consider whether certain statements could be possible, e.g. "The population in 1965 was 8000." Simply looking at the chart suggests that this is false, but this will highlight to students that they don't actually know anything about the population in between the recorded years: the line graph shows trends and it's only possible to make estimates of intermediate values.

What do you think? 16.3B

In question 1, students should recognise that the data is continuous and so a line graph is more suitable. There are many situations that could be represented by question 2, and students should consider several possible contexts. They could be stretched to add scales to the axes and give more details in each context.

Practice 16.3C

1 Students should be able to articulate why their bars should touch in this diagram whereas they didn't when drawing bar charts.

2 Watch out for students trying to tally the data too quickly and missing out pieces of information. They should be encouraged to work methodically to ensure that this doesn't happen. Students could compare their frequency diagrams: they may have used different scales so their answers will vary.

What do you think? 16.3C

These questions require students to think more deeply about why certain charts are more appropriate than others for displaying grouped quantitative data. In question 3, students should notice that certain values, e.g. 15, can be included in more than one row of the table. Students could draw a better table, and be able to explain why they can't complete it accurately from what Ed has done.

Consolidate – do you need more?

These questions revisit the key points covered in the previous exercises, ensuring that students can draw and interpret different types of charts.

Stretch – can you deepen your learning?

1 Students should draw the two-way table including the columns for bus and total, and then use the information in the question to complete it before drawing the multiple bar chart. As in the earlier example, students could be prompted to consider how their chart would change if they'd chosen to put a different category on the *x*-axis.

3 reinforces the idea that certain information can be *seen* by just looking at a chart whereas other information needs to be worked out.

4 Students should consider the effect of the different inequality symbols used in the table. They should realise that they don't know which version of the table is the correct one as they don't have the data set.

Reflect

Students should explain their learning in their own words as this will clarify their thinking. They should be able to highlight the similarities and differences between different types of charts and know which is the most appropriate to use for given sets of data.

16.4 Comparing data (1)

Small steps

- Find and interpret the range
- Compare distributions using charts
- Identify misleading graphs

Rationale

This chapter provides opportunities for students to compare data and to appreciate how charts can be misleading by exaggerating scales, etc. In the first exercise, students find and interpret the range in a variety of contexts. If students are struggling with the calculations, they could be encouraged to use a calculator to ensure that the focus is on finding and interpreting the range rather than on addition and subtraction. The second exercise looks at comparing distributions by means of charts, and how sometimes charts can be misleading because they exaggerate differences in values. Students should be able to explain the key features of a graph that make it misleading and suggest improvements. They should be able to make correct comparisons by extracting data from charts.

Things to look out for

- Using the first and last numbers from a data set rather than the smallest and greatest to find the range
- Not working in the same units when calculating the range
- Describing the range as lying between two values rather than calculating it
- Making incorrect conclusions from misleading graphs

Interleaving opportunities

- Averages
- Converting units
- Four operations
- Pie charts
- Types of data

Models and representations

- Number lines are useful for ordering data and working out the difference between a pair of numbers.

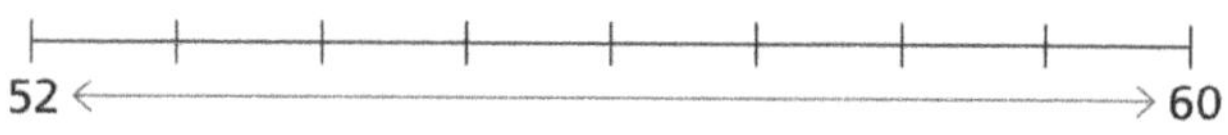

Behind the questions

Are you ready?

These two questions ensure that students have the essential prerequisite knowledge for this chapter. They should be able to identify the greatest and smallest numbers from a list of data and find the difference between any pair of numbers.

Practice 16.4A

1 Watch out for students calculating the range to be 280 g by just subtracting the first number from the last number. Students could be prompted to consider and explain this mistake even if they do not make it.

2 Students should know the difference between describing the spread of a data set and performing a calculation to work out the range.

4 highlights two common misconceptions. Students could be asked to demonstrate their understanding by calculating the correct value of the range as well as explain the mistakes made. Students could be stretched to consider the two possible answers to part b: one in grams and one in kilograms.

5 Watch out for students simply finding the difference between the two numbers given, rather than thinking about what they represent. A number line could be used for support, if needed.

6 A common mistake here is to not think carefully about the context and calculate the greatest time as 2 minutes and 76 seconds. If this happens, students should be encouraged to consider why their answer is technically correct, but why it is not given in the best way.

7 Watch out for students saying that the range is 5 or 3 because they just subtract the numbers attached to the year groups (i.e. calculating 6 – 1 or 5 – 2). They should think carefully about the context to understand why this would be incorrect.

What do you think? 16.4A

These two questions require students to think more deeply about the range. Question 1 revisits the idea of working in different units that was touched on in the previous exercise. Again, students could consider the two possible answers: one in litres and one in millilitres. Question 2 looks at how the range changes when more data is added. A number line can be used for support if necessary.

Practice 16.4B

1 Students should be able to articulate that a pie chart shows proportions rather than exact values. They should understand that as they don't know the exact number of cars in each car park, they can't make any comparisons about the numbers of cars. Instead, they can only compare the proportions of each colour.

2 Students should recognise that the scale on the vertical axis has been used to exaggerate the difference in the classes' scores. Students could be encouraged to draw the chart again using a different scale or construct a table to help them to answer part c.

3 Students should recognise that because they are told that both Emily and Seb have run 50 races, they can actually make comparisons about the numbers of times they've come in each position. However, they should also recognise that drawing Emily's pie chart bigger than Seb's is misleading because they've actually both run the same number of races.

What do you think? 16.4B

These questions require students to think about why a particular chart isn't appropriate for a data set and how using an incorrect chart can be misleading.

Consolidate – do you need more?

These questions revisit the key points covered in the previous exercises to ensure that students can calculate the range of a data set. If more practice is needed, students could be encouraged to work in pairs to create similar questions for each other to answer.

Stretch – can you deepen your learning?

1 Students should realise that the "pyramids" used here make it look as though the top of each chart is worth less than the bottom because its volume is less. They should also appreciate why a 3-D chart is difficult to interpret as it's hard to tell which part of the "bar" is lining up with the axis.

2 Watch out for students using trial and error. Instead, encourage them to form and solve an equation to find the value of x.

4 Students should reflect on their methods and the constraints they applied to x and y to find the answers.

Reflect

Students should explain their learning in their own words as this will clarify their thinking. They should be able to explain how to find the range of a data set and also identify the key features of a misleading graph. They could use examples to support their reasoning.

17.1 Using averages

Small steps

- Understand and use the mean, median and mode
- Choose the most appropriate average

Rationale

Students met both the median and the mean in Book 1, and here they are introduced to the mode. As well as finding the averages, there is a focus on comparing them and deciding which is/are most suitable for representing a distribution. This is best approached through discussion and comparison.

As well as the given sets of data, it is useful if students can collect their own data to analyse, or look at data from other sources that might spark interest. The range is not included in this chapter as students can mistakenly think that this is an average.

Things to look out for

- Mixing up the three averages – language association such as mode with "most" and median with "middle" can be helpful
- Not recognising the term "modal value" as meaning the mode – this will need explicit teaching
- Difficulty in choosing the most appropriate average – emphasise the purpose of averages to represent a set of data, and consider which do and which don't

Interleaving opportunities

- Link back to the charts drawn in Blocks 5 and 16, using any data already collected
- Revise giving results to numbers of decimal places and significant figures
- The averages are revisited in Chapter 17.3
- Outliers are formally discussed in Chapter 17.3
- You could work out percentage differences between the averages

Models and representations

- Organising data on a number line is a useful and quick visual representation. Highlighting the averages on this will show whether or not they are generally representative of the distribution.
- Likewise, bar charts, pie charts and other diagrams can be used. Any of these can be a very quick way of identifying a mode or a modal class.

Behind the questions

Are you ready?

These questions assess students' knowledge of the averages studied in Book 1 and will show if you need to remind students of these before moving on to the mode and comparing the averages.

Practice 17.1A

1 looks at the mode. Students need alerting to the fact that data sets can have more than one mode, or no mode at all.

2 to **4** provide practice at finding all three averages. You could introduce some comparison here if you wish.

5 looks at whether or not the averages need to belong to a data set. This is a good opportunity for discussion about whether they are representative. For example, is a mean of 1.8 children per family a useful measure?

6 and **8** require students to read from charts to find information.

7 introduces the term "modal value" as an alternative to "the mode". This is helpful when discussing the modal class.

What do you think? 17.1A

1 looks at the idea of a modal class and can be supplemented by discussion of any data students have already collected or used earlier in this book.

2 Students will need to be confident with simplifying fractions to identify the mode.

3 The ages are in different formats, so how best to compare them is a good discussion point.

Practice 17.1B

1 Many students will favour the mean as “the” average. The numbers here and the context provide a good example of how it can be distorted. You could introduce the term “outlier”, which is formally considered in Chapter 17.3.

2 The mean and the median are very similar here, in order to illustrate the fact there is not always an obvious answer.

3 illustrates that only the mode is possible for non-numerical data.

4 Here both the median and the mode represent the data well.

5 Students may not realise that $6\frac{1}{4}$ (the median) is not a shoe size.

6 looks at the way in which averages change when extra data is added. It is worth discussing suitable approaches, e.g. whether or not to start from scratch.

What do you think? 17.1B

These questions are designed to promote discussion. Use them in relation to students’ own data if possible.

Consolidate – do you need more?

This provides further fluency practice.

Stretch – can you deepen your learning?

1 is an opportunity to cement learning by using students’ choice of data. Secondary sources could be used as well as primary, e.g. sports results.

2 and **3** explore the effect on averages of changing the data. You could also look at the effect of adding an extreme value, in preparation for Chapter 17.3.

Reflect

Students reflect on the meaning and usefulness of each average.

17.2 Averages from tables (H)

Small steps

- Find the mean from an ungrouped frequency table (H)
- Find the mean from a grouped frequency table (H)

Rationale

You could use collected data to show that recording data in frequency tables and finding subtotals for each data value or group of values is much quicker and more reliable than adding up every individual data point from a list. The frequencies in the tables are quite high to emphasise this point.

These Higher steps are probably accessible for most students if calculators are used to prevent cognitive overload, or you could cover the use of a frequency table but not grouped data if appropriate for your class. If time is short, this could be omitted and covered in the review time at the end of Year 9 instead. The methods are covered with some practice questions in Block 17 of Book 3.

Things to look out for

- Students often divide by the number of rows instead of the total frequency – thinking about whether the calculated mean fits with the data highlights this error
- Incorrect multiplication by zero
- When using a table, students may think that the mode is the highest frequency rather than the data value with the highest frequency
- Students may be confused by tables presented horizontally – use both orientations in your examples and discuss the pros and cons of both ways of finding the averages
- Difficulty in finding midpoints – link this to a number line

Interleaving opportunities

- Link back to the charts drawn in Blocks 5 and 16, using any data already collected
- Revise giving results to numbers of decimal places and significant figures
- You could include discussion of finding or estimating the median and the mode/modal class from tables

Models and representations

- The frequency tables themselves are the main representation used in this chapter. You could compare them with the charts produced in Block 16 and look at how charts can be used to produce tables of values.

Behind the questions

Are you ready?

The first two questions check competency with the formula for the mean. Question 3 introduces the idea of getting an overall total from smaller sets of data, and the final question prepares students for finding the midpoints in grouped frequency tables.

Practice 17.2A

1 In this first question, a table is provided with the extra column needed for the subtotals. Check that students divide by 24 and not by 5.

2 Here students need to add in the last column themselves, and in question 3 they need to construct the table from scratch.

4 The table is presented horizontally. Discuss whether adding an extra row or rewriting the table vertically is preferable.

5 The table needs to be constructed from a chart. You could supplement this with other charts from earlier study.

6 includes some directed number arithmetic.

What do you think? 17.2A

1 looks at the most common misconceptions.

2 challenges students to find the median from a frequency table. This can be done by finding the cumulative frequencies, which can be demonstrated with or without the formal language. Students may need support to realise that the median is the value in the $\frac{n+1}{2}$th position rather than the nth position.

Practice 17.2B

1 A full table is provided for students to copy and complete. The fact that the sum of the midpoints serves no purpose is explicitly addressed.

2 does not provide students with the full structure of the table. It reminds students of the term "modal class".

3 presents the table horizontally. It promotes discussion of the likelihood of the midpoint being representative for smaller frequencies.

4 and **5** provide further practice, with the latter requiring interpretation from a chart.

What do you think? 17.2B

1 looks at the idea of equal distribution. You could compare this with data that students have collected themselves.

2 should promote an interesting discussion about to how to find a midpoint when the end point is unknown. This could be linked to a discussion of outliers.

Consolidate – do you need more?

Further fluency practice is provided if needed.

Stretch – can you deepen your learning?

1 is a useful investigation that is designed to show that the smaller the groups, the more likely the estimate is to be reliable.

2 The expectation is to use linear interpolation, e.g. the median might be found $\frac{2}{3}$ of the way through a particular interval.

Reflect

Students are encouraged to reflect on what's the same and what's different about the ungrouped and grouped tables.

17.3 Comparing data (2)

Small steps

- Identify outliers
- Compare distributions using averages and the range

Rationale

Identification of outliers is done by eye here rather than by using a formula, with a focus on how they affect the sizes of the mean and the range. Students should be encouraged to think about whether outliers are extreme values that may or may not be included, or recording errors that should definitely be removed or corrected.

The second part of the chapter builds on Chapter 16.4, this time using measures of location and the range to compare two or more distributions. A sophisticated analysis is not required: a statement comparing an average and another comparing a range are sufficient. You could develop this further by also looking at charts and describing the general shapes of the distributions, if desired.

Things to look out for

- Treating any value slightly away from the majority as an outlier, rather than just an extreme value
- Only looking for outliers at one end of a distribution instead of at both ends
- Thinking that a larger range is better than a smaller one, not realising that this implies inconsistency
- Always thinking that a larger average is better than a smaller one, which is not the case for data such as the time taken to complete a task
- Students may initially find making comparisons difficult, and this will need some modelling and discussion

Interleaving opportunities

- The averages used in the previous two chapters are revisited, as is the range from Chapter 16.4
- Link back to scatter graphs and consider outlying points

Models and representations

- Outliers are often obvious from diagrammatic representations of data. The use of a number line with dots to show each data point is particularly useful.

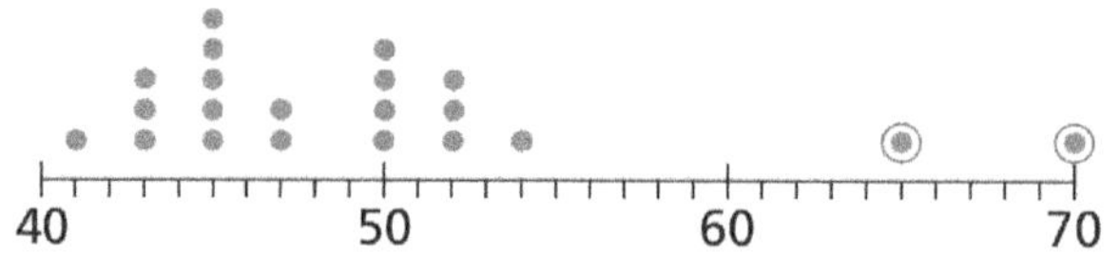

Behind the questions

Are you ready?

These fluency questions assess students' knowledge of finding the three averages and the range, which are the prerequisites for this chapter.

Practice 17.3A

1 and **2** The outlier is obvious in each case and has a dramatic effect on both the mean and the range.

3 introduces the idea of an item of data that has been recorded incorrectly.

4 Again the outlier is obvious, although some students may wish to consider November too. It is worth discussing that, although atypical, these values are an important part of overall performance in this context.

5, **6** and **8** provide further discussion points.

7 uses the number line model. You could ask students to construct similar models for the data in the earlier questions.

What do you think? 17.3A

1 requires working backwards to find the total from the mean. Some students may need support here.

2 to **4** are open discussion questions. Students should be made aware that statistics is not always an exact science, and interpretations can be subjective.

Practice 17.3B

1 asks explicitly for the two comparisons.

2 also requires two comparisons. Students may make only one comparison unless prompted.

3 and **4** Students need to perform the calculations before making comparisons.

5 looks at removing outliers.

6 generalises to more than two sets of data.

What do you think? 17.3B

This question ties both parts of the chapter together, as the range is severely affected by outliers.

Consolidate – do you need more?

This provides extra practice and further points for discussion, if required.

Stretch – can you deepen your learning?

1 There are plenty of examples of frequency polygons online. Encourage students to consider how they show the "shape" of the data.

2 This is a suggested context for investigation, but students could look at something different if you think they will find it more interesting.

Reflect

These very open reflections could start as discussions before students record their own thoughts.